T0331678

Newtonian
Mechanics for
Undergraduates

Essential Textbooks in Physics

ISSN: 2059-7630

Published

Essential Textbooks in Physics

Newtonian Mechanics for Undergraduates

Vijay Tymms
Imperial College London, UK

World Scientific

NEW JERSEY · LONDON · SINGAPORE · BEIJING · SHANGHAI · HONG KONG · TAIPEI · CHENNAI · TOKYO

Published by

World Scientific Publishing (UK) Ltd.

57 Shelton Street, Covent Garden, London WC2H 9HE

Head office: 5 Toh Tuck Link, Singapore 596224

USA office: 27 Warren Street, Suite 401-402, Hackensack, NJ 07601

Library of Congress Cataloging-in-Publication Data
Names: Tymms, Vijay, author.
Title: Newtonian mechanics for undergraduates / Vijay Tymms, Imperial College
    London, UK.
Description: New Jersey : World Scientific, [2016] | Series: Essential
    textbooks in physics | Includes bibliographical references.
Identifiers: LCCN 2015030925| ISBN 9781786340078 (UK) (hc : alk. paper) |
    ISBN 9781786340085 (pbk : alk. paper)
Subjects: LCSH: Mechanics--Textbooks.
Classification: LCC QC127 .T85 2016 | DDC 531--dc23
LC record available at http://lccn.loc.gov/2015030925

British Library Cataloguing-in-Publication Data
A catalogue record for this book is available from the British Library.

In-house Editors: Mary Simpson/Dipasri Sardar

Typeset by Stallion Press
Email: enquiries@stallionpress.com

Printed in Singapore

Contents

1

Overview

1.1. Introduction

I have been teaching physics for 16 years, starting with secondary school teaching, then later university lecturing where I taught the first year mechanics lecture course at Imperial College for four years from 2010–2014. Teaching this course has been one of the most enjoyable parts of my career thus far, giving me an opportunity to reinspect some of the most fundamental concepts in the discipline for delivery to a demanding (though appreciative) audience, complete with multiple demonstrations plus interesting problems and puzzles. During these years I developed and refined a set of comprehensive course notes tailored for the students I was teaching. This textbook is an adaptation of the notes, altered to appeal to a broader audience.

1.2. Why This Book is Needed

School syllabuses are in a state of constant flux. The breadth and depth of core physics and mathematics curricula taught in schools varies a little from year to year and a lot from generation to generation. So while well-established subjects in physics remain the same, the level of knowledge and understanding of students that enter university to study the discipline varies. This means that lecturers have to constantly update their courses to suit their target audiences and make the transition from A-level to degree as smooth as possible.

Although there are already many mechanics textbooks out there, there is a need for producing up-to-date reference material to match the level of development of the target audience. Essentially, textbooks quickly become out of date and there will always be a need for new ones. This particular one is designed to be in line with the level of physics and mathematics that contemporary school leavers ready to start a physics or physics-related degree will have.

1.3.   Who Will Benefit From This Book?

The lecture course that led to the creation of this book was designed specifically for first year physics undergraduates at Imperial College and as such the direct target audience of this textbook are students making the transition from school to university.

The book should also appeal to advanced A-level students unsatisfied with the level they have reached, and especially those who are considering studying physics or physics-related subjects beyond school. It contains some A-level material that is delivered at university level of presentation and should strengthen such students' understanding while also providing a smooth introduction to subtopics beyond the syllabus.

A-level physics teachers and first year university lecturers should also find the book useful; as well as the basic subject matter, in-depth examples and problems, there are also suggestions as to basic demonstrations that can easily be recreated in the classroom at minimal expense.

1.4.   Assumed Prior Knowledge

Regarding mathematics, all the content that can feature in a standard A-level mathematics syllabus is assumed knowledge throughout the text. Differential and integral calculus plus logarithms and exponents are used from the outset, with a gentle introduction to good practice in the use of integral calculus outlined early on. Knowledge of vectors is also essential, with scalar products being used from Chapter 5 and vector products from Chapter 14.

Regarding physics content, this course text goes from the ground up — i.e. in terms of classical mechanics everything starts from the beginning, though occasionally in examples and discussions some other subdisciplines of physics are invoked with a GCSE level of knowledge being assumed.

1.5. Structure and Topics

Readers of this text no doubt come from a broad range of backgrounds and have covered a wide range of high school-level syllabuses. This means that some readers will have a lot more knowledge of mechanics than others already. This course will start at a relatively basic level from the beginning and become quite advanced by the end. This means that all students will find some of the course to be revision, but it will vary from person to person. No one will find the whole of the course to be revision, and readers' understanding of familiar concepts will be further enhanced by the course in all cases.

An important thing to note about this textbook is that none of the material is redundant for any student taking a physics degree — every topic, subtopic, equation and example is relevant to the journey towards understanding physics. Whether you are aiming to excel at theory, experiment or computation, whether your interest lies in quantum field theory, atmospheric physics or cosmology, all of the material within these pages is relevant and will be beneficial on your route to becoming an expert.

The textbook follows a reasonably traditional route though is split into shorter chapters than most, which are of variable length. It starts with an overview, defining and categorising the most important quantities in the discipline, i.e. displacement, velocity, acceleration, force and mass. The next four chapters expand on this and bring in Newton's three laws of motion. Chapter 6 then briefly brings in linear momentum before introducing work and energy for the first time, leading to a much more in-depth view of momentum and potential energy. The second half of the book starts with motion on a curved path (not necessarily circular motion, though that is very much part of it) leading to simple harmonic motion and gravitation. Chapters 13 to 18 deal with the dynamics of rotating objects, ending with a brief look at gyroscopic motion and precession.

Feedback for the Author

I am happy to hear from any readers so please contact me via the publishers if you have any comments, queries or criticisms.

Vijay Tymms, April 2015

2

Introductory Concepts

This chapter introduces the five most important quantities in classical mechanics, namely displacement, velocity, acceleration, force, and mass. They are the most important quantities because it is impossible to make progress in the subject without first having an appreciation of what these quantities mean and how they relate to each other. The chapter provides definitions of the quantities at a level suitable to undergraduates and provides a discussion on what they physically mean.

The chapter also provides an introduction to scalar and vector quantities, SI prefixes and highlights a sensible approach to introducing new quantities and units that will be used throughout the text.

2.1. Quantities, Units, and Coordinate Systems

2.1.1. *Scalar and Vector Quantities*

In physics most measured quantities can be expressed by either:

(1) A magnitude only. These are known as **scalar** quantities or simply **scalars**. Examples of scalar quantities are time, mass, energy, power and density.

Mathematical operations on scalar quantities are familiar, i.e. they add, subtract, multiply, and divide like normal numbers.

Scalar quantities do not usually require any special notation to denote that they are scalars.

(2) A magnitude *and* a direction. These are known as **vector** quantities or simply **vectors**. Examples of vector quantities encountered in classical mechanics are force, acceleration, velocity, momentum and torque.

2.1.2. *When Vectors Will Be Used and What Knowledge Will Be Assumed*

Mathematical operations on vector quantities are less familiar and more complicated than with scalar quantities but you will be expected to know some of them to get through this book. *Vector addition and splitting vectors into components* are absolutely essential and will appear from Chapter 5 onwards. *Vector dot products* will be required from a similar stage, and *vector cross products* will be used in detail from Chapter 14 onwards. You should already have some familiarity with some of these topics from high school mathematics and you should refer to your favourite mathematics textbook or other resource if unsure. *Vector calculus* will not be used in this book as it is seldom seen at school and is usually first met midway through the first year of university degrees in physics and applied mathematics in the UK. On occasions where a knowledge of vector calculus could be utilised to enhance understanding, notes are written in the text with optional further reading cited in order to allow the interested student to pursue the topic further.

2.1.3. *Vector Notation in Print and in Handwriting*

When writing vector quantities, it is essential to use some sort of special notation to denote the vector. In this textbook they will usually be denoted in **bold**.

In handwriting, the two most common and generally unambiguous conventions are either to draw an arrow above ($\vec{v}$) or a tilde or straight line below the symbol ($\underline{v}$). Just as inclusion of these accoutrements implies a quantity is a vector, an omission means the quantity is a scalar, even if missed by accident. One must therefore be careful, especially if the handwritten work is going to be studied by someone else.

That said, there are occasions where a vector quantity can effectively be treated as a scalar (for example, when dealing with velocity along a straight line and the use of + and − symbols is sufficient to denote a direction); provided the preamble to such work states matters clearly enough in these situations then a vector notation can be omitted.

2.1.4. *Knowing When a Quantity is Scalar or Vector*

When dealing with quantities in physics, it is usually important to know whether the quantity is a scalar or a vector. In this book this will be stated whenever a new quantity is introduced. If unsure, a good question to ask yourself is whether directions are required when adding parts of the same quantity together. For example, if two times in seconds are added together it does not make sense to ask what directions they have (you do not add five seconds north to seven seconds west) but if two displacements in metres are added then directions must be considered. Sometimes this will not be as easy as it sounds but asking oneself this question usually assists in developing an intuitive feel for what a physical quantity means.

2.1.5. *Units*

SI units will nearly always be used in this book. In classical mechanics the *base SI units* are the kilogram, the metre and the second and all other *derived SI units* will be made from these.

However, for long units of time as humans we are thoroughly familiar with minutes, hours, days, weeks and years rather than times stated in large numbers of seconds. For example, it does not assist intuition to state a time as 2.2×10^8 seconds when 7 years would be better understood by all. When speaking about physics it is always best to set up scenarios in as easily understood a manner as possible, so for long times the more intuitive non-decimal system will be used when necessary.

There are four other base SI units in physics: The ampere (electrical current), the kelvin (thermodynamic temperature), the mole (amount of substance), and the less frequently encountered candela (luminous intensity) but they do not appear in purely classical mechanics. If this sounds surprising think of any derived SI unit that belongs within the discipline and reduce it to base units — you will find that anything solely within classical mechanics uses kilograms, metres, and seconds and any quantity which does use other units is part of other subdisciplines of physics.

2.1.6.   *Standard SI Prefixes*

The SI prefixes in Table 2.1 will be used throughout the text. If unsure then refer back to this table but as all physicists use these abbreviations as part of their vocabulary it is well worth memorising them if you do not know them already.

The milli- to atto- prefixes are sometimes referred to as *diminishing prefixes* and the kilo- to exa- as *magnifying prefixes*. Standard form for large numbers will of course be used as well.

Note that some of these prefixes are used more than others and some are used liberally for certain quantities but not for others. For example, a mass of 7.0×10^3 kg would rarely be referred to as 7.0 Mg by most physicists despite this notation being perfectly correct. This sort of thing is partly due to a logical decision to avoid confusion with other symbols (the "M" usually means "mega" for 10^6 but in handwriting could be confused for a lower case "m" for "metre") and partly due to convention or rather what is currently

Table 2.1: Standard SI prefixes

Multiplier	Prefix	Abbreviation
10^{18}	Exa	E
10^{15}	Peta	P
10^{12}	Tera	T
10^{9}	Giga	G
10^{6}	Mega	M
10^{3}	Kilo	K
10^{-3}	Milli	m
10^{-6}	Micro	μ
10^{-9}	Nano	n
10^{-12}	Pico	p
10^{-15}	Femto	f
10^{-18}	Atto	a

fashionable within the physics community. Whenever a new quantity is introduced, a short note on prefix notation and alternative units is provided.

2.1.7. *Coordinate Systems*

The Cartesian coordinate system (x, y, and z orthogonal axes) will no doubt be the most familiar coordinate system to most readers and will be the most often used system in this book. *Plane polar* or *spherical polar* coordinate systems are more natural for certain concepts involving spherically symmetries and will be referred to occasionally. Other coordinate systems will rarely be mentioned.

Although it may seem a little dogmatic on occasion to stick with Cartesian coordinates, this is deliberate — the purpose of the book is to enhance the reader's understanding of Newtonian mechanics and as such there are as few distractions as possible to deviate from this understanding. Sometimes, however, bringing in a different point of view (such as by presenting a physical scenario in an alternative coordinate system) can indeed enhance understanding and this is done when necessary.

2.2.   Time, Displacement, Velocity, and Acceleration

2.2.1.   *Time*

Time is assumed to be linear and universal. It is a scalar quantity with SI units of seconds (s). The second is one of the three *base units* in classical mechanics. The range of times that one might realistically encounter in physics range from 10^{-18} s, i.e. 1 attosecond (the sort of time scale for electrons to redistribute themselves during chemical bonding), to 10^{17} s (the approximate age of the universe). You may also have heard of the notion of the *Planck time* — a duration of approximately 10^{-43} seconds which is, according to some theories, the shortest possible quantum of time. For practical purposes, however, the shortest time we can currently measure is of the order of a few attoseconds.

Regarding the SI prefixes, all of the diminishing SI prefixes are in common usage for time so you will see ms, μs, ns, ps, fs, and as in use but you will never see the magnifying prefixes used as people will use standard form or the more normal human systems for recording longer times.

2.2.2.   *What is Meant By "Time is Linear and Universal" and Some Musing on Time Travel?*

No formal definition of time is provided, as it is assumed that all readers will be comfortable with what it is.[1] This is not really the place for a philosophical discussion on the fundamental nature of time, but let us provoke some thought by stating that when it is said that time is linear it is meant that it passes at a rate of 1 second per second (or, mathematically written, $\frac{dt}{dt}$ = exactly 1 (unity)) and

---

[1]Newton himself used words to this effect when mentioning time in some of his classic treatises.

that when it is said that it is universal it means that this is the case for:

- All observers,
- In all coordinate systems,
- At all points in space,
- At all points in time.

If time were not linear for any observer then they would essentially be a time traveller of some description:

- If an object found itself in a situation where $\frac{dt}{dt} > 1$ then it would be travelling forward in time faster than normal — i.e. it would be moving into the future,
- If $0 < \frac{dt}{dt} < 1$ then the object would be travelling forward in time but slower than normal,
- If $\frac{dt}{dt} = 0$ then an object would be "frozen" in time,
- and if $\frac{dt}{dt} < 0$ then an object would be travelling backwards in time.

This is good fun to think about, and lends itself well to science fiction. Further on in your physics career when learning about Einstein's theories of special relativity and general relativity you will be able to re-inspect these ideas and discuss whether the idea of non-linear, non-universal time is a possibility.

2.2.3. *Displacement*

The displacement of an object is defined as the distance and direction from its initial position, $x_{initial}$ to its final position, x_{final} — i.e.

$$x = x_{final} - x_{initial}. \tag{2.1}$$

Displacement is a vector quantity with SI units of metres (m).

Measurable displacements can range from 10^{-15} m $= 1$ fm for the linear dimension of an elementary particle (e.g. an electron) to 10^{26} m for the distance to the farthest galaxy, which can be thought of as an upper limit. As with time, there is a proposed fundamental quantum of displacement known as *Planck length*, which is well below the limits of human measurement and theorised to be of the order of $\sim 10^{-35}$ metres.

The scalar equivalent of displacement is **distance** and is also measured in metres. However, they are not the same thing; if you walked in a circle of perimeter 10 m and came back to where you started, your distance travelled would be 10 m but your total displacement would be zero.

The metre is the second of the three base units in classical mechanics. The SI prefixes in prevalent usage with metres range from kilo to femto with all the variants in between. It is relatively unusual to use Mm or anything higher. The Earth has a radius of approximately 6,400 km and most physicists will speak of it this way; hardly anyone refers to it as 6.4 Mm. Larger distances are on astronomical scales and standard form will often be used, or more convenient units such as the astronomical unit (AU — the distance from the Earth to the Sun) for distances within the reaches of our solar system and light years for larger distances still. For small distances the Angstrom — 10^{-10} m $= 100\,pm$ — is considered an old-fashioned unit but is still in common usage as it is handy for discussing atomic sizes.

2.2.4. *Velocity*

If a point of origin is defined and an object moves such that its displacement relative to this point of origin changes, then the object's velocity, v, relative to the origin is defined as its *rate of change of displacement (with time).*[2] In equation form this is written:

$$v = \frac{dx}{dt} \quad \text{or} \quad v = \dot{x} \text{ (pronounced "x dot").} \tag{2.2}$$

---

[2]Usually in classical mechanics, and, indeed, in many branches of physics "rate of change" implies rate of change with time (as opposed to anything else), hence the parentheses.

Both notations are equally valid, though in this book the former (Leibniz) notation will usually be used.

Velocity is also a vector quantity with a direction in the same direction as the infinitesimal change in displacement relative to the origin. It has SI units of metres per second ($\mathrm{ms^{-1}}$).

Velocities can range from approximately $10^{-9}\,\mathrm{ms^{-1}}$ for the relative motion between tectonic plates to $3.00 \times 10^8\,\mathrm{ms^{-1}}$ for the speed of light in a vacuum, which is an upper limit.

The SI prefixes for velocities in general use range from kilo to micro. The speed of light would rarely be stated as $300\,M\mathrm{ms^{-1}}$. For small velocities, other more intuitive units are often used; $10^{-9}\,\mathrm{ms^{-1}}$ is approximately equal to $3\,\mathrm{cm/year}$ which is the same as $30\,\mathrm{km}$ per million years which gives more of a feel for the distances objects move on geological time scales.

It is worth bearing in mind that the above definition is for an *instantaneous* velocity. Sometimes an *average* velocity, $\langle v \rangle$, is more useful, and is defined mathematically by:

$$\langle v \rangle = \frac{x_{final} - x_{initial}}{t_{final} - t_{initial}}. \tag{2.3}$$

This would be consistent with the definition of velocity as "displacement over time", which is redundant and overly simplistic once a rudimentary understanding of calculus has been achieved.

The scalar equivalent of velocity is an object's **speed** and is also measured in $\mathrm{ms^{-1}}$. If an object turns a corner or travels in a circle then it is possible for it to travel at constant speed, but not constant velocity as its direction is changing. An average speed would be given by the early school equation "distance over time".

As a final note here, bear in mind that an object's velocity is dependent on its reference frame — i.e. how fast an object moves depends on how fast an observer moves. For example a person may walk at $1\,\mathrm{ms^{-1}}$ relative to a stationary observer but will be stationary relative to a person walking alongside him at the same speed.

This reference frame dependence will be referred to many times in this book.

2.2.5. *Acceleration*

Acceleration, a, is defined as *rate of change of velocity (with time)*. This can be written in three equivalent ways in equation form:

$$a = \frac{dv}{dt} = \frac{d^2 x}{dt^2} = \ddot{x} \text{ (pronounced "x double dot").} \qquad (2.4)$$

Acceleration is a vector in the same direction as the infinitesimal change in velocity. It has SI units of metres per second squared (ms^{-2}).

Accelerating objects do not necessarily speed up or slow down. If an object's speed is constant but the direction of its velocity is changing then it is accelerating. An example would be a car driving round a corner at constant speed. Linear (i.e. straight line) acceleration is investigated in detail in Chapter 3 and curved path acceleration in Chapter 11.

The *acceleration due to gravity* has a value of approximately $9.8\,\text{ms}^{-2}$ downwards (i.e. towards the centre of the Earth); most everyday accelerations (that humans experience) are smaller than this. If a human being experiences an acceleration greater than this they are likely to be uncomfortable (sometimes deliberately so in something like a roller coaster). The greatest accelerations recorded are for electrons in particle accelerators and can reach $10^{27}\,\text{ms}^{-2}$.

SI prefixes for accelerations in general use tend to range from kilo to micro. For the large accelerations that charged particles experience, standard form is usually used. Often for human scale accelerations, fractions of g are quoted (e.g. the acceleration due to gravity on the Moon is $0.17\,g$ and the maximum linear acceleration of a Formula 1 car is about $1.5\,g$).

Another useful formula for acceleration follows from the *chain rule* of differential calculus. It can be seen that $\frac{dv}{dt} = \frac{dx}{dt} \cdot \frac{dv}{dx}$ and so acceleration can also be written:

$$a = v\frac{dv}{dx}. \tag{2.5}$$

Note that:

(i) This is *not* a valid definition for acceleration, rather it is simply another formula for acceleration. The definition is given by Equation 2.4 alone.

(ii) $\frac{dv}{dx}$, the rate of change of velocity with distance (note the vector notation for the v but not the x) is a vector quantity[3] with the direction in the same direction as the *infinitesimal change in velocity*. Is has units of ms^{-1} per metre (ms^{-1}/m). (This could of course be written as s^{-1} and perhaps strictly should be, but the former — although less elegant — conveys more information. For example, consider stating that a car increases its speed at a rate of $0.1\,ms^{-1}$ per metre. It is clear that for every metre the car moves its velocity increases by $0.1\,ms^{-1}$. But if the statement is merely that the car increases its speed at $0.1\,s^{-1}$ it would rightly confuse most people.)

(iii) That $v = \frac{dx}{dt}$ is the magnitude of the velocity (i.e. the speed) of the particle and is scalar.

As with velocity, Equations 2.4 and 2.5 refer to an instantaneous acceleration, when sometimes an average acceleration will be required. This will of course be given by:

$$\langle a \rangle = \frac{v_{final} - v_{initial}}{t_{final} - t_{initial}}. \tag{2.6}$$

---

[3]Firstly, note that as this is a rate of change of a quantity with something other than time it is absolutely necessary to say so. Secondly, note that it is a rate of change with *distance* not displacement — a scalar not a vector quantity. In calculus it is only possible to have a rate of change of quantity with respect to a scalar.

2.3.   Force, Mass (and Acceleration)

Force and mass are the most fundamental quantities in classical mechanics, and among the most important in physics. That said, the definitions for both may not seem as rigorous as for others (such as velocity), and indeed may vary from one source to another. Definitions used in this book are those that are suitable within the confines of this book and the physics addressed within. However, on occasion, alternative definitions are outlined when they enhance understanding.

2.3.1.   *Mass*

At school, and in elementary science textbooks, mass is sometimes loosely described as the amount of "stuff" in an object, or, occasionally, using slightly more sophisticated language, the "quantity of matter" in an object. This definition makes some sense: Everyday objects are made of atoms, which are in turn made of protons, neutrons and electrons, which are present roughly in equal quantities. As electrons are relatively light compared to the other two nuclear particles (approximately 1837 times lighter) one can say that heavy objects contain more protons and neutrons than light objects. This can be helpful. Consider the old conundrum: "which is heavier, a tonne of feathers or a tonne of gold?". Readers of a textbook at this level are of course unlikely to be tricked by this, but now consider the particles: The "amount of stuff" definition means that the number of protons plus neutrons in the feathers will be roughly equal to the number of protons plus neutrons in the gold — i.e. the number of basic mass-making particles in each is the same. Note that if our level of knowledge of what makes up matter only extends to atoms, and we thought of these as the indivisible chunks, then this would no longer work as the number of atoms in the gold will be significantly fewer than in the feathers (which are largely comprised of carbon atoms).

There is nothing wrong with the "amount of stuff" definition of mass provided it is used within the confines of explaining phenomena like the one described above. However, for classical mechanics a more fitting definition, and the one that will be invoked most often in this book, is that mass is an object's *reluctance to undergo acceleration*. This is also sometimes referred to as an object's *inertia*. What this means is that the heavier an object is, the more difficult it is to make it change its velocity, i.e. speed up, slow down or change direction.

Alternative definitions of mass will be seen in Sections 13.2 on gravitational field strength and 9.6 on mass–energy equivalence. None of these definitions is necessarily right or wrong, but rather they are useful within certain confines. Some discussion will be provided on the links between the definitions.

Mass is a scalar quantity with SI units of kilograms (kg). The kilogram is the third and final base unit in classical mechanics. Masses can range from approximately 10^{-30} kg for an electron to 10^{55} kg which is one estimate for the mass of the universe (though the "reluctance to accelerate" definition ceases to make sense when considering the whole universe as one single mass).

Regarding SI prefixes, the kilogram can cause confusion as the base SI unit already contains the kilo prefix but it is very definitely the kilogram and not the gram that is the base SI unit. For diminutive prefixes, milli, micro and nano are all used and below this physicists tend to either use standard form or other units. In particle physics for particle masses the electron volt (eV) is used. This will not be defined here but it is good for you to be aware that it is a unit of mass and not anything to do with electricity at this stage. Magnifying prefixes for mass in SI units are rarely used, with most scientists either using standard form or often tonnes (where 1 tonne = 1000 kg).[4] This is

---

[4]Note that 1 tonne is a metric tonne and is different from 1 ton which is sometimes known as an imperial ton and is approximately 1016 kg. Getting the spelling correct in writing is essential here as is using the full terminology in speech. Stick to kilograms to avoid confusion.

sensible as the potential to confuse the rarely used Mg with the often used mg in handwriting is clear. Oddly enough, kilotonnes, megatonnes, and gigatonnes are all commonly used units today.

2.3.2. *Force*

The most intuitive, and possibly most useful, definition of a force is any influence that causes an object to either:

(i) Speed up,
(ii) Slow down,
(iii) Change direction, or,
(iv) Change shape (deform).

As (i), (ii) and (iii) refer to changes in velocity (i.e. accelerations), an almost equivalent definition is to say a force is an influence that causes a *free* body to accelerate. A free body is an object that is not constrained by its surroundings — i.e. it is in a space without walls.

Force is a vector quantity with SI units of newtons (N). The newton is a derived unit and can be expressed in base units as kgms^{-2}. Everyday forces range from a few mN to a few kN; a convenient mental aid is that an apple on Earth weighs approximately 1 N. The gravitational force between the Sun and the Earth is approximately 10^{36} N.

All the standard SI prefixes are commonly seen in use with the newton.

Chapters 4–7 cover forces in detail, starting from Newton's laws of motion.[5]

---

[5]Note the use of the capital N when referring the person and the abbreviation for the unit, but the lower case n when writing out the unit in full.

2.3.3. *Relating Force, Mass, and Acceleration*

If a resultant external force (i.e. the sum of all the forces on an object), $\boldsymbol{F}$, is imposed on a mass m then the object will accelerate according to

$$\boldsymbol{F} = m\boldsymbol{a}. \tag{2.7}$$

The force and the acceleration are in the same direction. This is well verified by experiment under everyday conditions and is the fundamental equation underpinning all of classical mechanics. Chapter 4 will show how this equation is derived from an even more fundamental statement, Newton's second law of motion, and will highlight the circumstances under which it is valid.

2.3.4. **F = ma** *as a Cause-to-Effect Ratio and Other Examples in Physics*

Equation 2.7 could also be written (in a simplified form without vector notation) as:

$$a = \frac{F}{m}, \tag{2.8}$$

where F is the magnitude of the force and a the magnitude of the acceleration.

This is an example of a relationship in physics where a "cause-to-effect ratio" equals a "resistance to the effect". Here, the force *causes* the acceleration and the ratio of the two equals the mass — the resistance to the effect (acceleration) occurring.

There are some other examples of this type of relationship that you know and will meet during a Physics degree. One other example is seen in this course initially in Section 14.4 which is the rotational equivalent of Equation 2.8. Two other examples in physics are:

• In thermal physics, a temperature difference across a material causes a flow of heat through the material. The rate of flow of heat is dependent on the *thermal resistivity* of the material and the relevant equation is *thermal resistivity* $= \frac{temperature\ gradient\ (cause)}{rate\ of\ flow\ of\ heat\ (effect)}$. This is known as Fourier's law of heat conduction.

- You have probably already met Ohm's law in electronics where $resistance = \frac{potential\ difference}{current}$.

2.3.5. *Watch out for Careless Alternative Definitions*

Note that none of the three quantities in $\boldsymbol{F} = m\boldsymbol{a}$ have been defined using the equation itself. It is tempting and common to do so but can lead to circular definitions if used without due care (i.e. mass is defined in terms of force and thence force in terms of mass). Unfortunately, such circular definitions can occasionally be seen in some published texts.

2.3.6. *Definitions of the Second, Metre, and Kilogram*

Note that it has been assumed that all readers have an intuitive feel for how long a second lasts, how far a metre is and how heavy a kilogram mass feels. Notice also that the second, metre and kilogram themselves have not actually been precisely defined. This is strange in a sense, as nearly all derived units that are introduced in this book will have definitions provided. The reason for this is that though the definitions of the three base units are interesting, and in fact form a subject in their own right, the definitions are complicated and esoteric and do not really add anything to the understanding of classical mechanics. You are encouraged to read up on them elsewhere for your own interest, but please appreciate the definitions are not necessary for your understanding of the subject at hand.

3

1D Motion

This chapter looks at the physics of the motion of a particle constrained to move along one dimension (chosen to be the x-axis) only. It starts by deriving some familiar formulae — the equations for constant acceleration — by use of calculus, and goes on to look at more general systems where the acceleration is not constant.

As this is the first time integral calculus is used in this book, Section 3.1 goes through the procedure of setting up and solving an integral for a practical problem in great detail — more detail than you will ever usually see for this kind of problem. The purpose of this is to enhance understanding of the use of mathematics and encourage good habits in written work.

As this section deals with motion in 1D only, for the most part the vector notation need not be included; it is sufficient to specify one direction as positive and the other as negative without any risk of confusion.

3.1. The Equations for Constant Acceleration

3.1.1. Setting up the Basic Situation

Consider a particle of mass m moving with an initial velocity u with a constant applied resultant external force F as shown in Figure 3.1. No other information is provided on the situation — what causes the force and the nature of the particle is neither important nor relevant to the ensuing discussion.

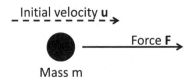

Figure 3.1:   **A particle of mass m travelling at speed u under a constant force F.**

The velocity at any other time is denoted by v. The particle's starting position is defined to be $x = 0$ when $t = 0$ and as x for any time afterward.

Applying $F = ma$ and using the definition of acceleration (Equation 2.4) to the particle gives:

$$\frac{dv}{dt} = \frac{F}{m} = a. \tag{3.1}$$

3.1.1.1.   *Finding v as a function of t, with a thorough and careful introduction to use of integral calculus in mechanics*

Because F and m are both constant, a is also constant. This *first-order ordinary differential equation* (ODE) is easy to solve by direct integration.

Although the mathematics is actually rather trivial for this first example, only the steps towards setting up this integral will be explained slowly and explicitly.

As a is a constant it can conveniently be written $\frac{dv}{dt} = a$.

In the Leibniz notation for calculus, derivatives are written in the same way as fractions — in fact, they are not fractions so they cannot be treated as such. Therefore, a naïve and incorrect mathematical step would be to write:

$$dv = a \cdot dt. \qquad \text{(Mathematical faux pas 3.1)}$$

It is important to realise that this equation should be considered sloppy and incorrect mathematics. Unfortunately, you will see many physicists treat differential equations this way; please resist the urge to join them.

While the above is wrong, to render it correct is simple. All that needs to be done is to place an integral sign on both sides of the equation as follows:

$$\int dv = \int a \cdot dt \qquad (3.2)$$

This is perfectly legitimate and correct mathematics. Essentially, what has happened is that the equation $\frac{dv}{dt} = a$ has had both sides integrated with respect to time. The dt terms on the left-hand side cancel — on this occasion it *is* acceptable to treat the equation as if it were a fraction. It is fine to go straight from Equation 3.1 to Equation 3.2 — the intermediate part must never be written as a separate line.

There are now two approaches to finding a final solution:

(i) Solving using an indefinite integral:

Equation 3.2 is written as an indefinite integral (i.e. there are no limits) and is easily solved to give:

$$v = at + constant. \qquad (3.3)$$

The information has been provided that when $t = 0$ then the velocity is u so substituting these values in gives $u = a.0 + constant$, hence the constant of integration is simply u and the complete solution to the problem is given by:

$$v(t) = u + at, \qquad (3.4)$$

where $v(t)$ implies that the velocity is a function of time. This is a standard type of notation that will be used throughout this book.

(ii) Solving using a definite integral:

Equation 3.2 is an indefinite integral. It can be written as a completely general definite integral as follows:

$$\int_{v_{initial}}^{v_{final}} dv = a \int_{t_{initial}}^{t_{final}} dt. \qquad (3.5)$$

Remember that the convention is that the final value taken by the variable is the upper limit of the integral and the initial value is the lower limit. The a has been taken outside of the integral as it is not a function of time.

For this particular problem the integral might therefore be written:

$$\int_u^v dv = a \int_0^t dt. \qquad \text{(Mathematical faux pas 3.2)}$$

As with Mathematical faux pas 3.1, this step is sloppy and incorrect, and unfortunately (as before) it is all too common to see physicists write this sort of thing. But, fortunately, as before, it is easy to correct. However, it must firstly be understood *why* it is wrong. The reason is that v and t are used both as *variables* when they appear as dv and dt in the integral *and* as the final values on the limits of their integrals. A variable cannot be given the same symbol as a final value.

The way to get round this is to give the variable quantity a different symbol to the final value, so, for example, Mathematical faux pas 3.2 could be properly written $\int_u^v d\phi = a \int_0^t d\psi$, where ϕ and ψ are newly introduced variables which fleetingly exist for the purposes of solving the equation. Variables of this kind are known as *dummy variables* (or sometimes "bound" or "dead" variables).

Rather than introduce completely new symbols, it is usually preferable and legitimate to simply change the variable to a primed quantity, thus Mathematical faux pas 3.2 can be properly written:

$$\int_u^v dv' = a \int_0^t dt', \qquad (3.6)$$

where v' and t' are pronounced "v prime" and "t prime". This solves to give:

$$v - u = at, \text{ i.e. } v(t) = u + at. \qquad \text{(Equation 3.4 reproduced)}$$

Whether you choose to solve problems of this kind by definite integration using limits or by indefinite integration followed by finding a constant is a matter of personal preference. In this author's experience, most undergraduates are more comfortable using indefinite integration. Definite integrals are usually employed in this book as they tend to be more elegant and involve fewer mathematical steps.

Having gone through a rather painstaking look at the mathematics, the analysis of the physics is fairly straightforward and the equation itself will be familiar to most readers. It is of the form $y = mx + c$ — i.e. it shows that $v(t)$ forms a straight line graph with gradient and intercept dependent on u and a as represented in Figure 3.2.

3.1.2. Finding x as a Function of t

Equation 3.4 can be written using the definition of velocity (Equation 2.2) to form another first-order ODE as follows:

$$\frac{dx}{dt} = u + at, \tag{3.7}$$

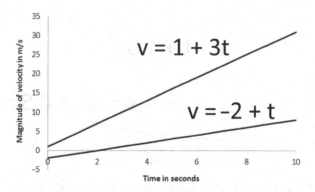

Figure 3.2: $v(t)$ for a constant acceleration for two different sets of parameters over 10 seconds. The higher line is for an acceleration of $+3\,\mathrm{ms}^{-2}$ and an initial velocity of $+1\,\mathrm{ms}^{-1}$; the lower is for $+1\,\mathrm{ms}^{-2}$ and an initial velocity of $-2\,\mathrm{ms}^{-1}$. A negative acceleration would be represented by a negative gradient, i.e. a downward slope.

which can be integrated with the parameters in this problem:

$$\int_0^x dx' = \int_0^t (u + at')dt'. \tag{3.8}$$

This is solved to give the familiar:

$$x(t) = ut + \frac{1}{2}at^2. \tag{3.9}$$

Equation 3.7 is of the form $y = ax^2 + bx + c$ with $c = 0$ — i.e. it shows that the graph of $x(t)$ takes the form of a parabola with the exact shape again dependent on the constants u and a as shown in Figure 3.3.

3.1.3. *Finding v as a Function of x*

If, instead of taking the original definition of velocity the alternative Equation 2.5 is used in conjunction with Equation 3.1, then the following first-order ODE is derived:

$$v\frac{dv}{dx} = a. \tag{3.10}$$

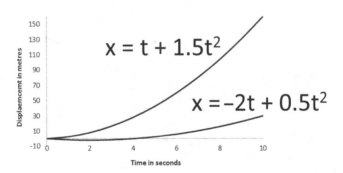

Figure 3.3: $x(t)$ for a constant acceleration for the same two sets of parameters as in Figure 3.2. In both cases, the curves are parabolas (the lower curve has the smaller acceleration and the less pronounced curve). The initial downturn on the lower curve indicates an initial negative velocity. A negative acceleration would be indicated with a downward sloping parabola.

This can also be solved using the parameters specific to this problem by integrating both sides with respect to x:

$$\int_u^v v'.dv' = a \int_0^x dx', \tag{3.11}$$

giving (after solving and rearranging):

$$v(x) = (2ax + u^2)^{\frac{1}{2}}, \tag{3.12}$$

which is again familiar (though written in a different format to the usual "$v^2 = u^2 + 2ax$"). It has been written in this format as it is more mathematically elegant to write the equation as a function of the dependent variable alone.

Note that the equation shows that when $x = 0$ then $v = u$ — i.e. $v(x = 0) = u$ as it should, and the form of the graph is shown in Figure 3.4.

3.1.4. *Two More Equations*

Equations 3.4, 3.9, and 3.12 cover all that is needed for constant acceleration. Two more equations can be derived by algebraic substitution (an exercise for the reader). They can be useful but do not

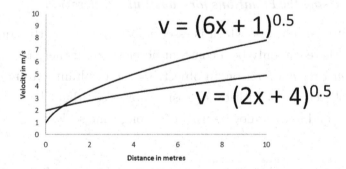

Figure 3.4: $v(x)$ **for a constant acceleration for the same two sets of parameters as in Figures 3.1 and 3.2 over a distance of 10 m. Note that the shape of the curves are that of a** $y = x^2$ **type parabola but rotated by 90° and truncated at the vertical axis. Mathematically it shows the inverse function of a parabola.**

provide any additional information:

$$x(v, t) = vt - \frac{1}{2}at^2, \tag{3.13}$$

$$x(v, t) = \frac{1}{2}(u + v)t. \tag{3.14}$$

If you covered these equations at school then you probably remember them anyway.[1] It is indeed important for any physicist to know these five equations; more important, however, is a knowledge and understanding of how to derive them, what they mean and when they are useful.

Being able to derive the equations using calculus is not the only possible method but it is an essential tool for the physicist and one which will be utilised many times in this textbook and one that you will require throughout your life in physics. It is recommended that you learn, practice and understand how to go through the proofs now as this will assist in your understanding of more complicated situations as the book progresses.

3.1.5.   *Using the Equations for Constant Acceleration*

In Newtonian physics, which is essentially governed by the equation $a = \frac{F}{m}$, there can only be a constant acceleration if there is constant *resultant external force* on an object.[2] If the resultant external force on an object varies with time, position or speed then the equations are not valid and cannot be used over long time scales.

---

[1]You may well know them as the "suvat" equations. While it is perfectly permissible to refer to them as such, be aware that this is relatively recent terminology which you will not find in either old textbooks or in the vocabulary of more senior physicists.

[2]If an object has varying mass *and* force, a situation can be engineered where acceleration would also be constant (though it is not as simple as having a varying force and making the mass vary at the same rate so the ratio of force to mass never alters). Section 4.4 revisits this idea.

Situations where the equations *are* valid are:

- In space away from any atmosphere or gravitational attractions so the total force is near negligible;
- Near the surfaces of astronomical bodies with no atmosphere or above their atmosphere where the gravitational field strength is approximately constant;
- For charged, free (i.e. not contained within an atom) subatomic particles and ions in uniform electric fields (where an individual particle is not undergoing frequent collisions);
- For free electrons under a constant potential difference in a conductor or semiconductor (that this happens to be true is very non-intuitive);
- FOR an object falling towards a planet with an atmosphere before the speed gets too large. On Earth, constant acceleration of $g = 9.8\,ms^{-2}$ is good for compact heavy objects for their first few seconds of free fall. The faster they get, the bigger the drag force. Once the drag force is a significant fraction of g, the net force will no longer be constant. The lighter the object and the greater its surface area, the less free fall time will be under constant acceleration;
- For any situation over a small enough time scale. "Small enough" in this case depends on how rapidly the acceleration is varying, and how accurately the system under analysis must be analysed. If, for example, measurements on the system need to be accurate to 1% and an acceleration varies by less than 0.9% over 5 seconds, then it would be fair to use the equations for constant acceleration for that 5 second duration. But if a 0.5% accuracy were required then they could not be used.

What this means in the universal scheme of physics is that the equations are only valid for a tiny minority of naturally occurring systems, but as we are humans that spend most of our time near the surface of our planet we often see objects falling downwards in free fall and for heavy objects we observe "constant g" accelerations all the time in our macroscopic existence.

3.2.   Time-Dependent Forces

Sometimes it may be possible to express a variable force on an object as a function of time alone. If this is the case then the force can be written mathematically as $f(t)$ and the corresponding ODE would be written:

$$m\frac{dv}{dt} = f(t). \tag{3.15}$$

Provided $f(t)$ is an integrable function with respect to time and initial conditions (or conditions at any point) are known exactly then $v(t)$ can be found. Integrating again (if possible) will allow the value of $x(t)$ to be found.

Example 3.1. An object initially at rest is pushed along the x-axis by a force given by $f(t) = kt$ where k is a positive constant. Find (a) the speed and (b) the total distance moved as a function of time.

The force in this case is one which starts at zero and then increases linearly with time. As the force varies, so does the acceleration and thus the equations for constant acceleration cannot be used. Instead we need to go to the basic definitions:

(a) Using $F = ma$ and the definition of acceleration it can be written $m\frac{dv}{dt} = kt$.

This is a first-order differential equation. Integrating both sides with respect to time gives the indefinite integral $\int m \cdot dv = \int kt \cdot dt$.

The question states that the object starts from rest so the initial velocity is zero and the corresponding definite integral with appropriate dummy variables is $m \int_0^v dv' = k \int_0^t t' \cdot dt'$.

This is simple to integrate and rearrange leading to:

$$v(t) = \frac{k}{2m}t^2.$$

Part (a) is now solved. But that is just the start: Having obtained a solution it is always a good idea to inspect it to see if it makes sense. There are a few ways to do this:

First of all, it is useful to check if the equation is dimensionally correct. The units of k are unfamiliar here, but as it is defined by $f(t) = kt$ where f is a force, they must be $\text{Ns}^{-1} = \text{kgms}^{-3}$ in base units. This means that the right-hand side of the equation must have units of $\frac{\text{kgms}^{-3}}{\text{kg}}\text{s}^2 = \text{ms}^{-1}$ which are the units of speed as they should be.

Second, it is useful to consider how raising or lowering a variable on the right-hand side of the equation would intuitively affect the dependent variable and decide whether the equation corroborates this intuition.

Considering time first, as t increases, v should also increase. It is not easy to guess how rapidly but there is a comparison with the equations for constant acceleration where $v = at$ for an object starting from rest — i.e. there is a linear increase. In this example, the force increases with time, so the speed should be expected to increase more rapidly than linearly and indeed a $v \propto t^2$ dependence is noted.

The larger the constant k is, the larger v would be expected to be and this is noted to be true.

And the larger the mass is, the more sluggish the object should be and indeed a $v \propto \frac{1}{m}$ dependence is seen.

(b) From this equation and the definition of velocity, $\frac{dx}{dt} = \frac{k}{2m}t^2$. The integral is therefore $\int_0^x dx' = \frac{k}{2m}\int_0^t t'^2 \cdot dt'$.

The expression for displacement is then:

$$x(t) = \frac{k}{6m}t^3.$$

The k and m dependences are consistent with those noted for the speed, and the time dependence is an $x \propto t^3$, which is again more pronounced than the constant acceleration version where $x \propto t^2$.

Though not required, $v(x)$can be found by eliminating t from the two equations (try it!) to render $v(x) = \sqrt[3]{\frac{9}{2} \cdot \frac{k}{m} x^{\frac{2}{3}}}$.

Verify for yourself that this is this dimensionally correct, and whether the equation seems intuitively plausible.

3.3.   Displacement-Dependent Forces

Sometimes a force can be a function of position alone. The most important example of such a force in this textbook is the gravitational force which is governed by the *inverse square law* and investigated in detail in Chapter 13. Here, let us look at the general case. If a force is a function of displacement only, it can be written as $f(x)$. The corresponding ODE *could* be written:

$$m\frac{dv}{dt} = f(x). \tag{3.16}$$

This is perfectly correct but is not in an integrable format — integrating both sides with respect to time renders:

$$\int m \cdot dv = \int f(x) \cdot dt. \tag{3.17}$$

This is not actually a mathematical faux pas but at the same time it is not particularly useful — it simply is not possible to integrate a function of one variable with respect to another. It is better to use Equation 2.5, i.e.:

$$mv\frac{dv}{dx} = f(x), \tag{3.18}$$

and provided $f(x)$ is an integrable function and initial conditions (or conditions at any point) are known exactly then $v(x)$ can be found. Integrating again (if possible) will allow the value of $x(t)$ to be found and $v(t)$ may be deduced by algebraic substitution. Note that in practice the sort of integrals that are produced at this second stage are non-trivial to solve.

An alternative approach is to write

$$m\frac{d^2x}{dt^2} = f(x), \tag{3.19}$$

and if a solution to this *second-order ODE* happens to be known then $x(t)$ can be found directly. An important example of this will be seen in Chapter 12 on *simple harmonic motion*.

Example 3.2. An object initially at rest is pushed along the x-axis by a force given by $f(x) = \frac{k}{x}$ where k is a positive constant. Find the speed as a function of displacement; the object starts at a distance ε from the origin.

Examples of this type of force are rare but this is not a completely contrived example. This would be the force on a point charge (i.e. a charge occupying an infinitesimally small volume such that its effective charge is concentrated at one point) in the vicinity of a long line of uniformly distributed charge of the same sign. Of course if two point charges are near each other, then the force varies as $\frac{1}{x^2}$, but one point charge near a line of charge varies as $\frac{1}{x}$. Incidentally, if a point charge is near a uniform sheet of charge, the force does not vary with distance at all. If the charges were of different signs then the forces would be attractive but have the same kind of dependence on distance; a similar result is seen for gravity.

Using $F = ma$ and the v, x way of expressing acceleration, $mv\frac{dv}{dx} = \frac{k}{x}$.

Integrating both sides with respect to x and putting in the necessary limits gives $m \int_0^v v' \cdot dv' = k \int_\varepsilon^x \frac{dx'}{x'}$. This integrates to give, before rearrangement:

$$\frac{1}{2}mv^2 = k\ln\frac{x}{\varepsilon}.$$

The left-hand side of this equation is a familiar term (though not yet encountered in this book) and is recognisable as the *kinetic energy* of the object. This term naturally appears when analysing a system this way; this will be revisited in more detail in Chapters 8 and 12.

The right-hand side of the equation shows that if ε were zero, the right-hand side would be a singularity, which makes sense as if ε were zero then the force would be divergent and therefore so would the velocity.

Rearranging gives:

$$v(x) = \sqrt{\frac{2k}{m} \ln\left(\frac{x}{\varepsilon}\right)}.$$

The velocity does get larger and tends to infinity with the distance but very slowly. This makes sense as the magnitude of the force becomes vanishingly small.

Note that the question does not ask for how x varies with time. The previous equation can be written:

$$\frac{dx}{dt} = \sqrt{\frac{2k}{m} \ln\left(\frac{x}{\varepsilon}\right)},$$

and integrating both sides with respect to time gives:

$$\int \frac{dx}{\sqrt{\ln\frac{x}{\varepsilon}}} = \sqrt{\frac{2k}{m}} \int dt.$$

The right-hand side is trivial but the left-hand side involves solving an integral of the form $\int \frac{dx}{\sqrt{\ln x}}$. This is an example of a function that is non-integrable — i.e. there is no mathematical function that exists that fits here; it is actually *impossible* to solve analytically. To make any progress, numerical approximations to a solution would have to be employed.

3.4. Velocity-Dependent Forces

If a force is a function of velocity only, $f(v)$, then the corresponding ODE would be written:

$$m\frac{dv}{dt} = f(v). \qquad (3.20)$$

Provided $\frac{1}{f(v)}$ is an integrable function and initial conditions (or conditions at any point) are known exactly then $v(t)$ can be found. Integrating again (if possible) will allow the value of $x(t)$ to be found.

Example 3.3. An object initially moving at speed u and starting at $x = 0$ moves in a resistive medium such that it is slowed by a force given by $f(v) = -kv$ where k is a positive constant. This is known as *Stokes' drag* and applies to objects moving at low speeds in a fluid where k depends on the fluid. Find (a) the speed and (b) the total distance moved as a function of time.

(a) Using $F = ma$ and the definition of acceleration, $m\frac{dv}{dt} = -kv$.
Rearranging and setting up the integral: $m \int_u^v \frac{dv'}{v'} = -k \int_0^t dt'$.
Integrating: $m \ln \frac{v}{u} = -kt$.
Rearranging: $v = ue^{-\frac{k}{m}t}$.
Notice that the velocity "decays" exponentially with time.[3]

(b) From this equation and the definition of velocity, $\frac{dx}{dt} = ue^{-\frac{k}{m}t}$.
Setting up the integral: $\int_0^x dx' = u \int_0^t e^{-\frac{k}{m}t'} \cdot dt'$.
Integrating and rearranging (try it!) gives $x(t) = \frac{um}{k}\left(1 - e^{-\frac{k}{m}t}\right)$.
This implies that the displacement tends to a maximum of $\frac{um}{k}$, a (possibly) counter-intuitive result. What this means is that although the object never stops moving its speed becomes vanishingly small and it nearly-but-not-quite reaches a final end point.

---

[3]The word "decay" is somewhat redolent of decomposition, and therefore the author prefers to simply say that the velocity falls exponentially.

Though not required, $v(x)$ can be found by either using Equation 2.5 or by eliminating t from the two equations (try it: It is a good exercise in the manipulation of logarithms and exponents) to render the linear relationship $v(x) = u - \frac{k}{m}x$.

3.5. More Complicated Forces

In general, forces will be a function of more than one parameter. For example, if there were a force in a gravitational field with air resistance and an exponentially decaying dependence with time then the differential equation describing its motion would be of the form $m\frac{d^2x}{dt^2} + b\frac{dx}{dt} + \frac{c}{x^2} = ke^{-\lambda t}$ where a, b, c, k, and λ are constants and solutions are more tricky to derive. Very often an exact analytical solution to the differential equation will not be possible and approximations or computational techniques will need to be invoked. However, what has been learned so far will cover a broad range of real life problems.

4

Newton's First and Second
Laws of Motion

This chapter introduces Newton's first and second laws of motion in their contemporary format. It explains when these laws are valid, shows how the newton is defined and gives some simple, qualitative examples of the use of Newton's second law in classical mechanics. Quantitative treatments follow in later chapters.

4.1. Newton's First Law of Motion

This can be stated as follows:

> **A body will move at a constant velocity unless a resultant external force acts upon it**

4.1.1. *The Law is Not Valid in Accelerating Reference Frames*

This law is valid only when the object is being observed in a reference frame that is not accelerating (i.e. an *inertial* reference frame). If the reference frame is accelerating (i.e. *non-inertial*) then Newton's first law will not hold. For example, an object that is stationary will appear to a non-inertial observer to be accelerating even though there is no force on it.

To give an everyday example which can cause people to misunderstand Newton's first law, consider yourself as the passenger in a car that goes around a sharp turn to the right at constant speed.

You feel as if you are flung out to the left. In reality, you are merely trying to travel in a straight line but the car accelerates during the turn and pushes you towards the right. During the turn, the people inside the car feel as if Newton's first law is not working as things seem to accelerate to the left without a force on them as the car is a non-inertial reference frame during the turn. Of course, to an external, inertial, observer on the road everything can easily be analysed in terms of normal Newtonian mechanics.

To deal with non-inertial systems properly either the use of so-called "fictitious forces" or general relativity are required. Both of these topics are outside of the scope of this textbook. They are usually met in optional courses in the third or fourth year of physics degree courses and are covered widely in the literature.

4.1.2.   *Nor is the Law Valid on Subatomic Scales*

Furthermore, the law is usually only valid at dimensions on the atomic scale and above. At subatomic scales the law becomes invalid and quantum mechanics (dealt with in detail in any physics degree) takes over. Indeed the idea of forces in quantum mechanics is seldom of much use in interpretation of either theoretical or experimental results at all. This most often needs to be borne in mind when considering subatomic particles, of which protons, neutrons and electrons are the most frequently considered in everyday physics. If a subatomic particle is a constituent part of an atom then it is said to be "bound" and the physics that governs its behaviour is entirely quantum mechanical. In this case, Newton's first law will not apply. If, however, the subatomic particle is not part of an atom and is comparatively far away from other particles then it is said to be "unbound" or "free" and then the particle can, for some situations, behave classically – i.e. it will obey Newton's first law. There is, of course, a lot more to be learned on this fascinating subject which forms a central part of any physicist's education. Prior to understanding quantum mechanics, however, a deep and intuitive understanding of classical mechanics must be developed.

In practice, Newton's first law of motion is used implicitly in understanding physical situations but very infrequently stated as a starting point in numerical problem solving.

4.2. Introducing Linear Momentum Before Stating Newton's Second Law

To introduce Newton's second law it is first necessary to introduce a new quantity: Linear momentum (often just abbreviated to momentum). Momentum will be revisited in detail in Chapter 7 but for now a definition and a few words on it will suffice.

The linear momentum p of a body of mass m travelling at velocity v is defined as the product of the mass and velocity, i.e.

$$p = mv. \tag{4.1}$$

Momentum is a vector quantity in the same direction as the body's velocity with SI units of $kgms^{-1}$ or Ns. It is fine to use either; though it is easier for now to see why the former unit might be used, the use of the Ns will become clearer when momentum is seen in more detail later along with the related quantity impulse. It may be better to stick with Ns when using SI prefixes to avoid confusion with altering the kg part. SI prefixes for momentum tend to range from kilo to micro with standard form used for values out of this range. For example, a housefly might have a magnitude of momentum of approximately $10\,\mu$Ns; a lorry on the motorway might have a magnitude of momentum of about $100\,k$Ns.

Newton himself referred to momentum as an object's "quantity of motion". A better explanation may be that momentum is a measure of how difficult it is to stop an object — i.e. reduce its velocity to zero. It is also essential to bear in mind that, just as with the velocity, an object's momentum is reference frame-dependent.

4.3.   Newton's Second Law of Motion

This can be stated as follows:

> **The rate of change of momentum of a body is directly proportional to the resultant external force that acts upon it.**

If the resultant external force is $\boldsymbol{F}_{RE}$ then the law can be mathematically expressed by:

$$\boldsymbol{F}_{RE} \propto \frac{d\boldsymbol{p}}{dt}. \tag{4.2}$$

(Note the $\boldsymbol{F}_{RE}$ is often written $\sum_i \boldsymbol{F}_i$ to signify the sum of all the separate forces on an object. The sum notation will not usually be used here.)

As with the first law, this law only holds for inertial frames of reference and for scales above the subatomic level. Furthermore as the speed of an object approaches the speed of light, the law requires modification as momentum is no longer directly proportional to velocity. This is dealt with in special relativity, which all readers will have heard of and some will have met formally. The first year of most physics degree courses has a course on the subject for those who are still unfamiliar with the concept.

Newton's first law was essentially deduced by Galileo in the mid-17th century and was modified, clarified and written up by Newton along with the third law of motion and the law of gravitation in *Principia* in 1687. For everyday situations (i.e. situations on a macroscopic scale and at velocities much slower than the speed of light) the laws are well verified by experiment, and other than within the realms explicitly stated no experiment has ever been done that contradicts these laws. The first two laws were often (and sometimes still are) stated to be fundamental — i.e. they are a basis with nothing deeper behind them to explain *why* they are true. It was not until the mid-20th century (some years after the advent of quantum mechanics) that physicists (notably Richard Feynman) grew to realise that Newton's laws of motion are a natural *consequence* of the laws of quantum mechanics (see Ogborn and Taylor, "Quantum physics explains Newton's laws of motion", Journal of Physics Education, January 2005 for a readable paper on this).

4.4. Derivation of $F = ma$ and the Definition of the Newton

Newton's second law in the format given in Equation 4.2 is certainly useful but still is not actually an equation (hence the quotation marks — an equation requires an equals sign and this is more properly referred to as a proportionality relation) and can be reworked into a more useful format. As the resultant external force is directly proportional to the rate of change of momentum, "Equation" 4.2 can be properly written as an equation as follows:

$$F_{RE} = k\frac{dp}{dt}, \qquad (4.3)$$

where k is a dimensionless constant. In SI units k is *chosen* to be unity — i.e. exactly equal to 1.[1] This gives:

$$F_{RE} = \frac{dp}{dt} = \frac{d(mv)}{dt} \qquad (4.4)$$

using the definition of momentum.

---

[1]Simply stating "k is chosen to be unity" may seem a little too easy and will rightly cause some minor consternation with a few readers and thus merits further thought: The point is that k is set to 1 in the SI system of units as this is one of the first things we meet in physics, and this actually leads to clumsy-looking constants appearing elsewhere in the subject as a result, notably in electromagnetism. If you look up the formal definition of the SI unit of electric current (the ampere), for example, you will note that the definition uses a force of magnitude $4\pi \times 10^{-7}$N, which looks really awkward. If the ampere had been defined first, the definition of the newton might be the quantity with the unusual constants tacked on the front.

Equation 4.4 can subsequently be expanded using the product rule of differential calculus to give:

$$\boldsymbol{F}_{RE} = m\frac{d\boldsymbol{v}}{dt} + \boldsymbol{v}\frac{dm}{dt} = m\boldsymbol{a} + \boldsymbol{v}\frac{dm}{dt}, \tag{4.5}$$

where $\boldsymbol{a}$ is the body's acceleration, which is $\frac{dv}{dt}$ by definition.

The second term on the right-hand side of Equation 4.5 involves a rate of change of mass. Physics involving changing masses and requiring the use of this equation in full will be dealt with in Chapter 10. For many situations, however, the mass is constant, so the rate of change of mass is zero. Hence, for the special case of non-changing mass Equation 4.5 reduces to the familiar.

$$\boldsymbol{F}_{RE} = m\boldsymbol{a}. \tag{4.6}$$

This is the form in which Newton's second law of motion is most frequently (though rather casually) stated, and most often used. Much of this course will use the law in this form, and it is fine to start problems in mechanics with an obviously non-changing mass by writing "Using $F = ma\ldots$".

It is this equation that leads to the definition of the newton: It is defined to be *that force which causes an acceleration of exactly $1\,ms^{-2}$ to a mass of exactly $1\,kg$.*

4.5.   Simple $F = ma$ Examples for a Point Particle

This section looks at types of force and free body diagrams in some detail. First we inspect some simple examples of $F = ma$ for different situations:

4.5.1.   *No Velocity, Balanced Forces*

If an object is ever stationary with respect to the inertial reference frame then the sum of all the forces on the object *must* be zero. For example:

(a) If a book is at rest on a flat table then the weight must equal the contact force.

(b) If a book is at rest on an inclined table then the three forces on it (weight, contact force, friction) must add to give zero.

(c) If a ladder rests against a wall then the five forces (what are they?) on it must add to give zero. The leaning ladder problem if investigated in detail in Section 15.5.3.

4.5.2. *Constant Velocity, Balanced Forces*

If an object is ever moving at constant velocity with respect to the inertial reference frame then the sum of all the forces on the object *must* be zero. For example:

(a) A parachutist on a calm day falling at a steady speed must have their weight equalling the drag force.

(b) If a book slides down a table at constant speed then the same three forces on it (weight, contact force, friction) must add to give zero.

(c) If an object is dragged along the ground at constant velocity then the four forces on the object (tension in the rope doing the dragging, weight, contact force with the ground and friction with the ground) must add to give zero.

4.5.3. *Constant Acceleration, Unbalanced Forces*

If an object is ever moving at constant acceleration with respect to the inertial reference frame then the sum of all the forces on the object *must* be non-zero and *must* be constant. For example:

(a) An object dropped on the Moon has a constant weight only.

(b) An electron between two parallel uniformly charged plates has a Coulomb force only (actually it will also have a negligible weight on the Earth).

4.5.4. *Non-Constant Acceleration, Unbalanced Forces*

If an object is ever moving at non-constant acceleration with respect to the inertial reference frame then the sum of all the forces on the object *must* be non-zero and must be changing (either in direction or magnitude). Some examples were outlined in Sections 3.2–3.4. For example:

(a) An object falling into the Sun from a large distance away will have a single force on it (the gravitational force, i.e. the weight) which increases as the object gets closer.

(b) An object orbiting the Earth at fixed distance has a single force on it (the weight) which is constant in magnitude but is constantly changing direction (it always points towards the Earth's centre).

(c) Dropping a ball bearing into a liquid will cause the ball bearing to initially accelerate at g but approach a terminal velocity as the force varies with velocity.

4.5.5. *Force Implies Acceleration and Acceleration Implies Force; Deduction and Induction*

It was made clear in Section 2.3 that a resultant external force causes an acceleration and not the other way round. But as they are linearly correlated by $F = ma$ it means that:

(1) If we are ever unable to actually observe a system but are able to analyse all the forces on it and decide that they do not add to zero — i.e. there is a resultant external force present — then we can conclude that the system must be accelerating in the direction of the resultant with a magnitude given by $\frac{F}{m}$.

(2) If we make an observation of an object without any analysis of physics at all and observe that it is accelerating we can conclude

that it must have a resultant external force on it in the direction of the acceleration with a magnitude given by ma.

The first way of investigating a problem is sometimes referred to as a *deductive approach*, and the second way as an *inductive approach*. Sometimes when analysing problems it is necessary to take an approach which wholly uses one of these extremes but usually the physicist with some common sense will use a mixture of both approaches to get to the heart of a problem and fully understand the Physics.

4.6.   Alternative Statements of the Laws

The statements of the first two laws of motion in Sections 4.1 and 4.3 are suited to the style and scope of this book. There are many other ways of stating the laws and no two textbooks will have an identical wording. Some are more simplistic while some appear more complicated but they all essentially say the same thing. Expressing the second law in equation form is fine provided the symbols are all explained but beware of texts that simply state the law as $F = ma$. It should be clear that this is less general than the full statement involving rate of change of linear momentum and thus is not quite correct.

5

Types of Force and Free Body Diagrams

This chapter introduces the most important practice in the analysis of systems in classical mechanics: The use of the free body diagram. It provides a list (with an associated study of the physics) of the types of macroscopic everyday mechanical forces that we encounter with free body diagram examples where appropriate.

5.1. Free Body Diagrams

A free body diagram (or force diagram) is a picture of a mechanical system with vector arrows showing all the external forces on a single object within the system. The sum of the vectors gives the resultant external force on the object, and applying Newton's laws of motion can thereby allow us to understand how that object will behave. This is particularly applicable to solid, rigid objects but can be used with some success for fluids as well.

An accurate free body diagram uses arrows to depict the forces where the arrows:

(i) Show the direction of the forces,
(ii) Have labels to indicate the type of forces,
(iii) Have relative magnitudes suitable to indicate the relative magnitudes of the forces, and,
(iv) Have their origin at the point at which the force acts from.

Sometimes sketching a free body diagram can be rather trivial but sometimes it can be very tricky indeed. Identifying the types of force and adding the necessary labels is usually clear enough. Intuiting the relative magnitudes of the forces can be difficult, especially for dynamics situations with rotating and accelerating objects. However, drawing a free body diagram should always be the starting point for a physicist analysing a system in terms of the forces involved. Even if unsure about whether or not the initial diagram is correct, it requires the illustrator to think in detail about the physics of the problem. Often once the initial diagram is drawn and Newton's laws have been applied, it will become clear that the diagram has some subtle errors and thus refinements must be made. Only when the diagram is completely correct can the illustrator be said to fully understand the physics of a situation.

5.2. Types of Mechanical Force

You may have heard that there are only four types of force — gravitational, electromagnetic, strong nuclear and weak nuclear — and that of these, only the first two are of any importance outside of the nucleus of the atom. This is true, but on a macroscopic scale, with solid objects in particular, it is possible to divide the forces that arise as a result of gravitation combined with electromagnetic repulsion and attraction into sensible categories.

5.2.1. *Weight*

Weight, **W**, is defined as the force on an object due to the effect of a gravitational field. In a gravitational field of strength **g**, an object of mass m has a weight of

$$\boldsymbol{W} = m\boldsymbol{g}. \tag{5.1}$$

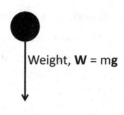

Figure 5.1: **Free body diagram showing an object's weight. Note that on the surface of a spherical planet the weight acts towards the centre of the planet, and if near the surface so the ground appears flat, acts perpendicular to the ground.**

Equation (5.1) implies that the weight is in the same direction as the field; near the surface of a planet an object's weight acts downwards and perpendicular to the ground. A free body diagram for an object in a gravitational field is shown in Figure 5.1.

The weight force is always drawn to act from the *centre of gravity* of the object, which is defined in detail in Chapter 15. For now, it is usually ok to simply draw the weight as acting from the centre of the object.

It is clear to us from experience exactly what happens to such an object — it accelerates downwards, and most readers will at some point at school have measured the acceleration to be $g \approx 10\,\mathrm{ms}^{-1}$ on the surface of the Earth. We can reach the same conclusion by analysing the physics: applying $F = ma$ to the object in Figure 5.1 gives $ma = mg$ as the weight is the sole force on the object. The masses cancel and the acceleration is always simply g for any object with only a weight force.

5.2.1.1. *A preferred way of stating "the masses cancel"*

Saying the "masses cancel" is mathematically correct but always sounds a bit glib. A more detailed way of explaining this is to say the increase in force caused by a greater mass is exactly balanced by the increased reluctance to accelerate.

Gravitational fields and the concept of weight will be revisited in Chapter 13.

5.2.1.2.   *Mass and weight in common parlance*

Most of us will spend the entirety of our lives within the confines of the Earth's gravitational field and all objects we handle are pulled downwards with a weight of magnitude mg. In speech we talk about how much objects weigh and when referring to how difficult they are lift upwards then this is just about acceptable terminology. Of course, when people refer to an object weighing a few grams, what they really mean is the object has a mass of a few grams and most physicists will reluctantly concede that any quest to change the way the English-speaking world uses mass and weight is a lost battle.

Another inaccuracy is to say that when an object rest on some scales, that the scales record the object's weight. Even if the scales were calibrated to deliver a reading in newtons rather than kilograms this would not quite be right. In fact, what the scales record is the magnitude of the normal contact force between the object and the scales.

5.2.2.   *Normal Contact Force*

If a solid object gets pushed by a solid surface then the force between the surface and the object is known as the **normal contact force** (usually given the symbol R or N). It acts perpendicular to the plane of the contact between the surfaces.

While we are all familiar with the notion of a contact force, and we more or less permanently experience one acting upon us, it is worth considering for a moment what it actually is. It is a macroscopic (i.e. large scale) phenomenon that follows as a result of atomic scale forces. To investigate the nature of a normal contact force in detail would require a view of the interaction of the atoms composing each surface at the interface. At this level the forces are electromagnetic in nature and what we consider as a contact force is actually a sum of the forces on the particles at the boundary between two objects.

Whenever a normal contact force on an object is present it is spread over the surface of the object so there really is no single *point* of contact, rather there is an *area* of contact with the force spread over the area. This can make life a little difficult when a free body diagram is required, as the arrow is supposed to originate from the point the force acts from. It is usually most helpful to draw a single arrow acting from the centre of the contact surface to represent the force; however, for certain dynamic situations this can be misleading so care must be taken.

Figure 5.2 shows the forces on an object in a lift for three situations.

Applying $F = ma$ to the object and using magnitudes only in Figure 5.2 gives:

(a) $ma = R_a - mg = 0$ i.e. $R_a = mg$.
(b) $ma = R_b - mg$ i.e. $a = \frac{R_b}{m} - g$ or $R_b = m(a + g)$.
(c) $ma = mg - R_c$ i.e. $a = g - \frac{R_c}{m}$ or $R_c = m(g - a)$.

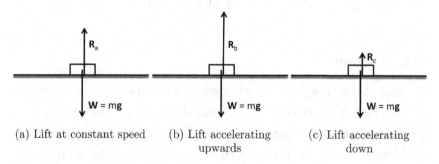

(a) Lift at constant speed   (b) Lift accelerating   (c) Lift accelerating
                                   upwards   down

Figure 5.2: **The normal contact force on an object resting on the floor of a lift (a) moving at constant velocity, (b) accelerating upwards, and (c) accelerating downwards.**

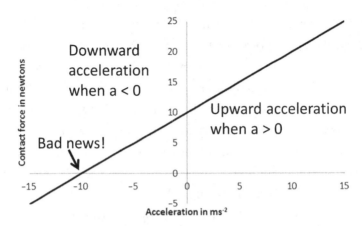

Figure 5.3:   Contact force vs. acceleration for a 1 kg object in a lift on Earth. The contact force increases linearly with acceleration. If the contact force is zero it means that the lift is accelerating downwards with a magnitude equal to the acceleration due to gravity alone. In practice this is likely to mean bad news as the lift has likely suffered some mechanical failure. The negative contact force indicated for accelerations of $<10\,\text{ms}^{-2}$ would correspond to the lift being pulled downwards "faster than g" with the objects either tied or glued to the floor so that it is pulled downwards.

What happens in case (c) if the downward acceleration of the lift equals the acceleration due to gravity?

A plot of the value of the contact force with lift acceleration is shown in Figure 5.3 for a mass of 1 kg and with $g = 10\,\text{ms}^{-2}$.

5.2.3. *Friction*

If a solid object moves — or feels a tendency to move — over a rough, solid surface it experiences a **frictional force**, F, that acts in the *opposite direction* to the motion, or the tendency towards motion. By a tendency to move, it is meant that something is trying to push the object in a certain direction but the friction is big enough to prevent the motion.

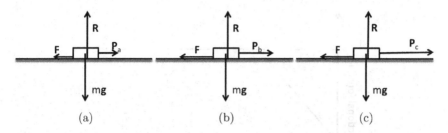

Figure 5.4: **Frictional force opposing the intended direction of motion for three different magnitudes of applied force. (a) Low applied force. (b) Applied force at threshold. (c) Applied force above threshold.**

If the object is stationary, resting on level ground, and a force of gradually increasing magnitude and constant direction acting parallel to the ground is applied, then:

(i) The frictional force *equals* the applied force until a certain *threshold force* is reached.

(ii) Once the threshold force is reached, and the applied force continues to increase in magnitude, the frictional force *stays at the same maximum value*. This is illustrated in Figure 5.4.

In Figure 5.4, for (a) the object will not accelerate, in (b) the object still will not accelerate but is *about to* and in (c) the object will accelerate at a rate $\frac{P_c - F}{m}$. An idealised graph of frictional force vs. applied force is shown in Figure 5.5a.

For everyday situations the frictional force between two objects are (loosely) governed by *Amonton's Laws of Friction*. These state:

• The threshold frictional force is directly proportional to the normal contact force between the two objects.

In equation form this is stated as:

$$F \propto R, \quad \text{i.e. } F = \mu R, \tag{5.2}$$

where μ is a constant known as the *coefficient of friction* between the materials that make up the two objects. It is a dimensionless

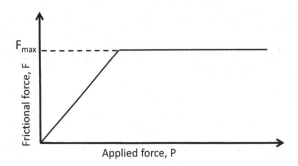

Figure 5.5a: Frictional force vs. applied force for an ideal object on a surface for a continually increasing applied force.

scalar with a value that depends on the *two* surfaces in contact. For materials on wet ice, μ can be lower than 0.1, while for some materials on rubber it can be as high as 1. (It is often mistakenly believed that 1 is an upper limit for the coefficient of friction but this is not the case as will be seen shortly.) When stating a coefficient of friction it is necessary to mention both materials — it makes no sense to state the coefficient of friction of a single material as it always depends on both the materials in contact.

- The frictional force is *independent of the area of contact* between the two objects.
- The coefficient of friction (and hence the maximum frictional force) is *independent of the sliding object's velocity.* When an object is sliding it will always experience this maximum frictional force.

5.2.3.1.   *How much can Amonton's laws be trusted?*

It is possible that you may feel these laws may seem at odds with everyday experience and indeed they are only correct to first order. For approximate calculations they work well but do not stand up to scrutiny when an accuracy of better than 10% is required. They are also useful in building understanding in simple situations for students learning classical mechanics as idealised problems with friction governed by Amonton's Laws are can be highly instructive.

5.2.3.2.   *Kinetic friction and static friction*

If an object is pushed with a steadily increasing force until it starts to slide, the pusher may note that the force required to keep the object moving is a little less than to start it moving. Try this now with a heavy object on a flat table. This is more than mere perception, it is indeed correct, and in fact the *coefficient of static friction* is always a little higher than the *coefficient of kinetic friction*. This means a more accurate representation of frictional force vs. applied force is shown in Figure 5.5b.

5.2.3.3.   *Example on static friction: the inclined plane problem*

It is absolutely essentially for any physicist to understand the behaviour of objects on an inclined plane (i.e. a slope) under gravity. Several examples of increasing sophistication will be seen in this book, starting here with a relatively simple one (not quite *the* simplest, which would be an inclined plane with no friction at all): an object at rest on an inclined plane.

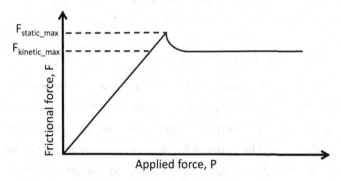

Figure 5.5b: A more realistic frictional force vs. applied force curve for a continually increasing applied force.

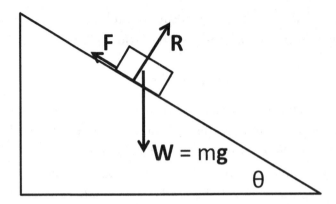

Figure 5.6:   Free body diagram for a stationary object on an inclined plane.

Consider an object at rest on a plane inclined at θ to the horizontal. The forces on it are its weight, the contact force by the plane, and the friction which "holds it in place". A free body diagram for the object is shown in Figure 5.6.

The system can be analysed by examining perpendicular force components. As the problem is set up in $2D$, this means we need to choose two axes at $90°$ to each other. There is a freedom of choice with selection of these axes — any orientation will work to give the same answer — but the most sensible for this situation are either:

(a) axes parallel and perpendicular to the *ground* or

(b) axes parallel and perpendicular to the *plane.*

5.2.3.4.   *Which axes to select?*

To make life easy in terms of the mathematics, it is best to choose axes where the least splitting of components will occur. In this situation, the weight acts perpendicular to the ground, the friction acts along the plane and the contact force acts perpendicular to the plane.

This means that if option (a) is selected the weight will not be split into components but the contact force and friction will be. But if (b) is selected only the weight will need splitting into components.

As option (b) involves just one force to be split into components, not two, this is going to provide the easiest mathematics.

5.2.3.5. *Solving*

So selecting option (b), the component of weight perpendicular to the plane is $mg \cos \theta$ and along the plane is $mg \sin \theta$. Hence:

as there is no acceleration perpendicular to the plane $R = mg \cos \theta$;
as there is no acceleration parallel to the plane $F = mg \sin \theta$.

As the plane angle increases, the contact force decreases and the frictional force increases. When the threshold force is reached the object will slip. This will happen at a certain critical angle, θ_{max} and it can be seen that this can determine the coefficient of static friction easily by $\mu = \tan \theta_{max}$ (can you see why?).

It is helpful to examine a plot of the forces on the block with angle. Figure 5.7 looks at the case for an object of mass $1 \, \text{kg}$ with $g = 10 \, \text{ms}^{-2}$.

The plots show that if $\mu < 1$ then the threshold friction will be reached when the tipping angle is less than $45°$. This is the case for most materials. If $\mu > 1$ then the tipping angle is greater than $45°$.

5.2.4. *Tension and Compression*

5.2.4.1. *Tension*

If a rope (or string or cord — in discussion of classical mechanics the words tend to be used interchangeably) is used to pull an object then the rope is said to be *in tension* and a rope under tension in classical physics will instantly transfer the applied force from one end of the rope to the other. In fact the force gets transferred at the speed of sound in the material that makes up the rope but for distances of metres this will be near enough instantaneous. Often when developing an understanding of concepts in physics it is necessary

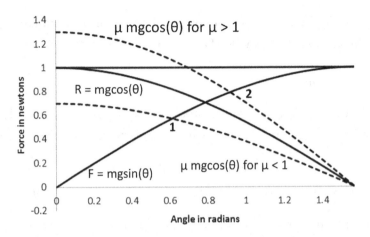

Figure 5.7: Weight, contact force and frictional force with angle for an object on an inclined plane. The value of $\mu mg \cos\theta$ for $\mu > 1$ and $\mu < 1$ are also shown. If an object rests on a plane which starts level and is gradually tipped, the frictional force increases. When the friction reaches this value of $\mu mg \cos\theta$ it will start to slide. These are marked as points 1 and 2.

to invoke an idealised rope that is both massless (termed "light") and inextensible. The tension for an idealised rope is uniform along the length of the rope. There is further discussion of the concept of tension when developing Newton's third law in the next chapter.

Consider, for example, a person dragging a sack along the ground with a force of 500 N. The person transmits the force to the rope via the friction force between hands and rope. This 500 N force then transmits itself to the sack (if a knot is tied this transmission of force will essentially be caused by a contact force between the rope and the sack). It is slightly inaccurate though perfectly acceptable to refer to the force on the person and the force on the sack as both being tension even though the former is a frictional force, and the latter a contact force.

As another example, consider a picture of mass m hanging symmetrically from a wall as shown in Figure 5.8.

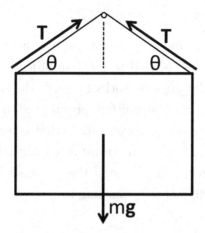

Figure 5.8: **A picture hanging from a wall with a light, inextensible string draped over a peg.**

Given that the picture is in equilibrium, horizontally the tensions balance and vertically the components of the tension add to equal the weight so that $2T \sin \theta = mg$, so $T = \frac{mg}{2 \sin \theta}$. As $T \propto \frac{1}{\sin \theta}$ for a narrow painting the tension will be relatively low, with a minimum value of $0.5\,mg$ for an infinitesimally narrow painting corresponding to the weight being equally supported by two strings. As the painting gets wider the tension increases so the sum of the upward components can balance the weight; as the width becomes very large the tension tends to infinity so it becomes impossible to support very wide paintings in this way. The wider the painting supported this way, the stronger the string needs to be.

5.2.4.2. *Tension can transmit a force around a corner*

Notice that the tension "bends around" the peg. This illustrates a general point provided there is no energy dissipated by friction at a bending point, the tension in ropes can be used to make an applied pulling force "turn a corner" (as happens in a pulley system for example).

5.2.4.3. *Compression*

If a pole (or stick or rod — again, these are interchangeable words
in classical mechanics) is used to push an object, it is said to be *in
compression* and the physics works in much the same way as with
tension, except that the transmitted force is used to push rather than
pull. The nature of solid materials means that is much more difficult
to get a compressive force to be transmitted around a corner than it
is for a tension force. This is one of the reasons why pulling a load
is usually more practical than pushing it.

5.2.5. **Upthrust**

5.2.5.1. *Density*

The density, ρ, of a material (solid, liquid or gas) is defined as its
mass, m, per unit volume, V, i.e.

$$\rho = \frac{m}{V}. \tag{5.3}$$

It is a scalar quantity with SI units of kgm^{-3}. Everyday densities
range from $1\,\text{kgm}^{-3}$ for air, to $19000\,\text{kgm}^{-3}$ for gold. Density is a
property of a *material* and in physics is referred to as an *intrinsic*
quantity. This can be contrasted with an object's mass which is specific to a certain object and is referred to as an *extrinsic* quantity.
The idea of intrinsic and extrinsic quantities is of great importance
in physics but is not of much value in pursuing discussion in classical
mechanics. You will probably meet the concept in most detail in the
study of thermodynamics at university level.

Usually, no SI prefixes are used for densities and whole numbers
are given, possibly with standard form for light gases. Another useful
unit is gcm^{-3}, where $1000\,\text{kgm}^{-3} \equiv 1\,\text{gcm}^{-3}$, so $\rho_{gold} \approx 19\,\text{gcm}^{-3}$.
As water has an approximate density of $1\,\text{gcm}^{-3}$, densities are often
given relative to water and are referred to as *specific densities*.

5.2.5.2. *The upthrust (buoyancy) force*

If a solid object is fully or partially submerged in a fluid (liquid or gas) then it will experience an *upthrust force,* U. The size of this force is given by Archimedes Principle which states that:

> **If a body is fully or partially submerged in a fluid, then the fluid exerts an upward force on the body equal to the weight of fluid displaced by the body.**

This means that an object less dense than a fluid will float in the fluid, and an object denser than a fluid will tend to sink in a fluid. Figure 5.9 shows the free body diagrams for a light object and a heavy object in a fluid.

5.2.6. **Drag Force**

If an object moves through a fluid, it experiences a **drag force** that opposes the motion. The size of the drag force depends principally on the velocity of the object, the fluid and the shape of the object in motion (specifically the frontal surface area). Stokes' drag was seen in Section 3.4, and *Newtonian drag* will be discussed in Section 10.7.

5.2.7. **Lift**

5.2.7.1. *Pressure*

Pressure, P, is defined as the magnitude of force, F, per unit area, A. Mathematically,

$$P = \frac{F}{A}. \tag{5.4}$$

It is a scalar quantity with SI units of pascals (Pa) (often written Nm^{-2} and equivalent to $kgm^{-1}s^{-2}$ in base units). Pressure is a measure of the "concentration" of a force. By pushing your thumb hard on a table you can create a pressure of about $10\,M$Pa and by lying down flat your weight can be spread to as low as $2\,k$Pa.

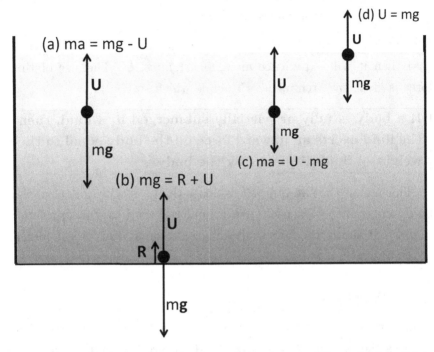

Figure 5.9: Free body diagrams for (a) a heavy object sinking (b) a heavy object at rest (c) a light object rising and (d) a light object at rest. In this case, "heavy" means denser than the fluid and "light" means less dense than the fluid.

Pressure was alluded to but not specifically mentioned in the discussion on normal contact forces earlier in the chapter.

5.2.7.2. *Lift force*

If a fluid flows past a certain asymmetric shape (e.g. a shark's fin or an aeroplane's wing) the pressure caused by the fluid flow on each side of the object will be uneven. This pressure difference (when multiplied by the area) gives a force perpendicular to the direction of fluid flow. The resulting force is called the lift and is significant.

The physics behind the lift force forms a subject in its own right and will not be addressed in this textbook.

6

Newton's Third Law of Motion

This chapter introduces Newton's third law in a contemporary format. It provides an insight into how Newton's third law is used to understand physical situations with examples of Newton's third law pairs and provides a warning about careless use of the law.

6.1. Newton's Third Law of Motion

The law can be stated as follows:

> If a body A exerts a force on a body B then body B exerts a force on body A that is
>
> (i) equal in magnitude
> (ii) in the opposite direction
> (iii) of the same type.

This law is *always* true. There are no known exceptions. Having said this there are some subtleties to the law that appear when analysing systems in quantum mechanics and electrodynamics when the use of forces is open to debate. This is addressed in a little more detail in Section 6.3.

Newton's third law of motion is of use in identifying the forces acting on a system. To analyse the corresponding motion of the system, Newton's second law of motion is used.

6.2. Newton's Third Law Pairs

Two bodies exerting a force on each other according to Newton's third law are referred to as "Newton's third law pairs". The following list explains how Newton's third law pairs exist for everyday physical situations and macroscopic forces. In some cases specific examples are given. Note that a diagram highlighting a Newton's third law pair is *not* a free body force diagram as it does not purport to show all external forces on a single object in the system.

6.2.1. *Type 1: Long Range Forces ("Action at a Distance")*

1. The gravitational force (weight)

We usually say that the Earth attracts falling objects but it is just as correct to say that falling objects attract the Earth. Newton's third law tells us the gravitational forces will be equal in magnitude and opposite in direction.

For example, consider an apple falling from a tree towards the centre of the Earth: The Earth exerts a gravitational force *on* the apple directed *from* the apple's centre *towards* the Earth's centre. Therefore, by Newton's third law the apple exerts a gravitational force of equal magnitude and opposite direction along the same axis *on* the Earth, as shown in Figure 6.1.

In this example, if the apple has a mass of 100 g, and a corresponding gravitational force on it (the weight) of 1 N, and the Earth has mass 6×10^{24} kg then from Newton's second law, an inertial observer will see the apple accelerate "down" at $10 \, \mathrm{ms}^{-2}$ and the Earth accelerate "up" at $1.7 \times 10^{-25} \, \mathrm{ms}^{-2}$. As $F = ma$, the relative magnitudes of the accelerations can be seen to be the inverse magnitude of the ratio of the masses: We can state that $m_{apple} a_{apple} = m_{Earth} a_{Earth}$, hence $\frac{a_{apple}}{a_{Earth}} = \frac{m_{Earth}}{m_{apple}}$.

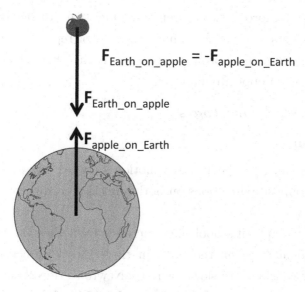

Figure 6.1: A Newton's third law pair for the gravitational force between an (exaggeratedly large) apple and the Earth. This is *not* a free body diagram as it does not show all the forces on one single object.

It can further be deduced that the ratio of velocities and displacements from the original positions are also in the same ratio (why?). How far will they both move if the apple falls from a height of 10 m?

2. **Electromagnetic forces of attraction and repulsion**

You will deal with these in detail at various stages of your physics degree course — most notably when studying electromagnetism — but it is worth remembering that Newton's third law applies to all these systems. Examples include:

(a) Electrostatic forces: Two charged particles attract or repel each other with identical forces in opposite directions;

(b) Magnetostatic forces: a permanent magnet that attracts a piece of iron feels an attractive force towards the iron of the same magnitude;

(c) Lorentz forces: two current-carrying wires attract (when the currents move in the same direction) or repel (when the currents run in opposite directions) each other with identical forces in opposite directions.

6.2.2. *Type 2: Contact Forces*

3. Friction

Friction opposes the relative motion between two objects and acts in opposite directions on both objects according to Newton's third law.

For example, if a book slides across a table then the table puts a frictional force *on* the book in the opposite direction to the velocity causing it to slow down. By Newton's third law, the book therefore puts a frictional force <u>on</u> the table of the same magnitude, but in the opposite direction — i.e. in the direction of the book's velocity (see Figure 6.2).

If the table is not fastened down, or on wheels, then this forward force will cause the table to accelerate in the same direction as the book's velocity. If this does not seem realistic, think of a person running and jumping onto a stationary skateboard causing the skateboard to move forwards.

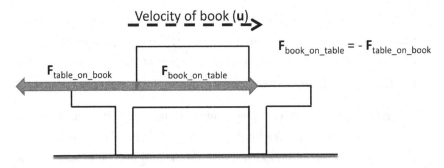

Figure 6.2:   **An example of a Newton's third law pair using the frictional force. Once again, this is *not* a free body diagram.**

4. The normal contact force

If one object pushes another, the push is called the "contact force" and by Newton's third law it does not matter which of the objects is responsible for the pushing — whatever push force is administered, the "non-pushing" object pushes back with a force of equal magnitude but in the opposite direction.

This may seem counter-intuitive: if a person strikes a punch bag, for example, it may seem that the person's fist does all the pushing. But if it is recalled that the fist *decelerates* on contact with the punch bag, it is clear from Newton's second law that there must be a resultant external force — the contact force — on it.

For example, consider a person jumping off the ground. They push down on the ground with a contact force which exceeds their weight which causes them to accelerate upwards (by Newton's second law). The contact force starts at the same value as the person's weight, rises, causing the person to accelerate upwards, then falls to zero as the person leaves the ground. By Newton's third law, the contact force *by* the person *on* the ground will be of equal magnitude and opposite direction (i.e. straight down). This is illustrated in Figure 6.3.

5. Tension

If a rope is in tension then a Newton's third law pair of forces exists. Imagine the rope is put in tension by two people, A and B, pulling on each end (as in a tug of war). Provided the rope does not slip from either person's hand, if the tension in the rope is T, each person experiences a force by the rope that is numerically equal to T but acts in opposite directions for both.

If the two people are on frictionless surfaces (or skateboards) then they will both accelerate towards each other with same force (and with a ratio of accelerations that is inversely proportional to the ratio of their masses, i.e. $\frac{a_A}{a_B} = \frac{m_B}{m_A}$, just as with the apple

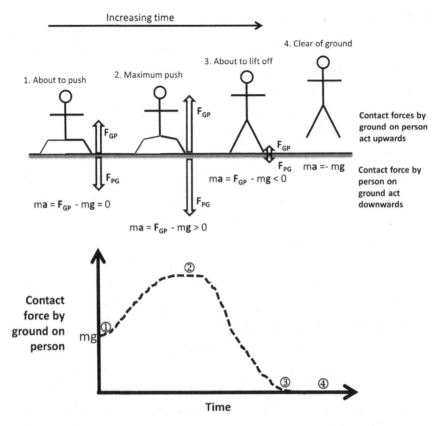

Figure 6.3: A schematic diagram of a person jumping showing the contact forces as a Newton's third law pair. These are *not* free body diagrams. The Newton's second law equations refer to the person. The plot shows the approximate form that the contact force will take over time.

and the Earth. It does not matter who does the pulling — the tensions will be the same for both.

Considering a rope made of segments of equal length

A more profound understanding of the nature of tension is given by considering a rope in tension, attached to a wall at one end and pulled with a force at the other. Imagine the rope to be made of segments of equal length and think of the force on each of these segments as shown in Figure 6.4a.

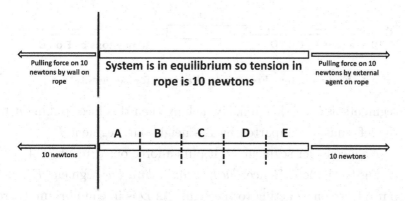

Figure 6.4a: A rope under tension can be thought of as a series of equal segments. In this case, the external agent pulls with a force of 10 N so the wall must pull on the rope with the same force for the system to be in equilibrium and the tension is also 10 N.

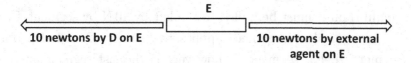

Figure 6.4b: Free body diagram for segment E of the rope.

Each of these segments must also be under a tension of 10 N — i.e. this force pulls each segment in both directions. This can be shown by recognising that if the whole rope is in equilibrium, then each segment must also be. The mechanics of each segment can be considered in turn.

Starting with segment E, consider its free body diagram (Figure 6.4b):

Because E is in equilibrium and it is pulled to the right with a force of 10 N by the external agent, there *must* be a force of 10 N pulling it to the right by Newton's *second* law.[1] If we consider the

---

[1] A very common, and completely erroneous, explanation at this point is to state the similar "Because E is in equilibrium and it is pulled to the right with a force of 10 N by the external agent, there *must* be a force of 10 N pulling it to the right by Newton's

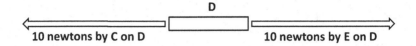

Figure 6.4c: Free body diagram for segment D.

segments as being like links in a chain then this force pulling it to
the left must be imparted by E's neighbour, segment D.

Now consider segment D: it puts a force on segment E of 10 N
acting to the left. Hence, *by Newton's third law* segment E must
put a force on D acting to the right. As D is in equilibrium, there
must be a force of 10 N pulling it to the left, imparted by segment
C. The free body diagram for D (Figure 6.4c) is thus:

For segment C a similar analysis applies: by Newton's third
law, D puts a force of 10 N on C to the right, and as it is in
equilibrium it must be pulled to the left by 10 N by segment B.

An equivalent argument applies to segment B and then for
segment A. For segment A, however, as the last segment in the
line, the force pulling it to the left is not by another segment, but
by the wall.

The wall is also in equilibrium, so the force by A on the wall
is 10 N to the right. As this force has ultimately come from the
external agent and passed down the segments this force can ultimately be said to be *by the external agent on the wall*, acting
through the rope as a tension.

Note that the choosing of five segments is arbitrary. Any number of segments would merit the same analysis and in Newtonian
mechanics it is often convenient to think of the number as being
infinite. In reality of course, matter is grainy and thinking of the
problem microscopically we would ultimately need to think of the
force on individual atoms.

---

third law". Simply changing one word has rendered the physics badly wrong. Note how
the third law is properly invoked in the next paragraph.

6.2.3. *Type 3: Fluid Pressure Difference Forces*

Newton's third law can also be applied to forces due to fluid pressure (for example, drag, upthrust and lift) but not in the same way as with action at a distance and contact forces. In these cases it is sensible to think about the pressure being caused by the rate of change of momentum of the molecules of atoms making up the fluid when they collide with the side of a solid object. The force on the molecule forms a Newton's third law pair with the force on the solid object. The mechanics of collisions (including molecular collisions) will be looked at in more detail in Chapter 10, though the subtler concepts of fluid pressure are not for this book.

6.3. Misuses and Apparent Paradoxes

A warning: Newton's third law is one of the most misunderstood and misinterpreted laws in science. As well as the conceptual difficulty of the law providing a source of confusion for students learning physics it is often invoked very casually by people outside of the field. Internet forums especially will occasionally provide "evidence" that the law has exceptions, occasionally with thought experiments to back up the argument. Sometimes it can be quite an intriguing exercise to try and find the flaw in the discussion.

6.3.1. *Action and Reaction*

One old-fashioned form of the law that is in common parlance is the well-known "every reaction has an equal and opposite reaction". Use of this form of the law is to be discouraged, both amongst physicists and also among the less initiated. Physicists should be aware that use of the word "action" to mean force is now considered out of date, and "action" in mechanics is now defined as something completely different to force. Amongst non-physicists, the law can be heard uttered as a way of saying "what goes around comes around", which is reasonable while it is merely a figure of speech but unfortunately can sometimes be heard to lend strength to arguments with an air of expertise, which is a feeble appeal to authority.

7

Linear Momentum

This chapter reintroduces linear momentum and shows how the conservation of linear momentum follows as a consequence of Newton's third law of motion. It looks at simple collisions in one and two dimensions.

7.1. Linear Momentum

Linear momentum was introduced and defined in Section 4.2 to be the product of a body's mass and velocity, i.e. $p = mv$ (Equation 4.6). It was necessary to introduce the quantity in order to effectively state Newton's second law of motion, which finally led to the equation $F_{RE} = ma$ (Equation 4.1) for a constant mass. However, a more general formula was given by Equation 4.4 and this is the more general equation format of Newton's second law, i.e. $F_{RE} = \frac{dp}{dt}$.

7.2. Change in Momentum: Impulse

Consider a resultant external force that acts on an object for a finite amount of time, t. If the initial momentum of the object is p_1 and the final momentum of the object is p_2 then, following Newton's second law, the change in momentum is given by:

$$\int_{p_1}^{p_2} dp = \int_0^t F_{RE} dt. \tag{7.1}$$

This change in momentum due to a force acting over a specific time interval is known as the *impulse*, I, of the resultant external force

73

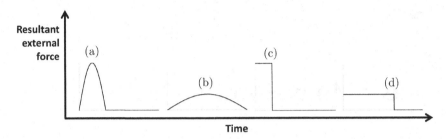

Figure 7.1:   **Force–time graphs with the same enclosed area and therefore the same change in momentum (impulse) applied to the object subjected to the force along one dimension. (a) shows a large force for a short time (like a sharp blow), (b) a small force for a large time (a soft blow with follow-through), (c) a large constant force for an short time and (d) a small constant force for a long time.**

and is hence given by:

$$I = p_2 - p_1 = \int_0^t F_{RE} dt. \qquad (7.2)$$

Impulse is a vector with the direction defined by the change in momentum and with the same SI units as momentum.

From Equation 7.2, it can be seen that the impulse on a body due to a force acting over a certain time will be given by the area under a force–time graph as definite integrals provide the area under a curve between the two limits. The plots in Figure 7.1 show four force–time graphs which would lead to roughly the same impulse supplied to the object the force acts on.

Impulse and force–time graphs are of particular use in the analysis of short bursts of non-constant force (e.g. hitting a golf ball).

7.3.   The Conservation of Linear Momentum

7.3.1.   *Proof of the Conservation of Momentum for a General Two Particle System*

Consider an isolated system (i.e. subject to no external forces) consisting solely of two bodies, A and B, which interact (i.e. exert a force on each other) over a time interval t. The following analysis is general and applies to any type of force over any time duration.

Considering body B only: The force by A on B is given by $\boldsymbol{F}_{AB}$ and the impulse acting on B will therefore be $\int_0^t \boldsymbol{F}_{AB} dt$.

This quantity represents the change in momentum of body B as a result of the interaction with A.

Considering body A only: The force by B on A is given by $\boldsymbol{F}_{BA}$ and the impulse acting on A will therefore be $\int_0^t \boldsymbol{F}_{BA} dt$.

This quantity represents the change in momentum of body A as a result of the interaction with B.

Hence, the change in linear momentum of the *whole system* is given by $\int_0^t \boldsymbol{F}_{AB} dt + \int_0^t \boldsymbol{F}_{BA} dt$.

By Newton's third law of motion, the force by A on B is equal and opposite to the force by B on A, or mathematically, $\boldsymbol{F}_{AB} = -\boldsymbol{F}_{BA}$.

So the expression for the *total change in momentum* becomes $\int_0^t (-\boldsymbol{F}_{BA}) dt + \int_0^t \boldsymbol{F}_{BA} dt = 0$.

What this implies is that:

> **For any two-body isolated system of two particles, the change in momentum of the particles is zero regardless of whatever interaction occurs for those particles**

Or in other words:

> **Provided no external forces act on a two particle system, the total linear momentum remains constant**

7.3.2. *Conservation of Momentum for an N-Particle System*

Consider an isolated system of N particles. Each one of these particles interacts with every other one. For each of these pairs of interactions, the same analysis as applied above applies. Hence, the general rule is:

> **Provided that no external forces act on a system the total linear momentum of the system remains constant**

This law is known as the **conservation of linear momentum**. Because it follows directly as a consequence of Newton's third law of motion, which has no known exceptions, it too has no known exceptions.

In fact, even though we have derived the conservation of momentum as a consequence of the third law, it can be thought of as a more fundamental concept outside of classical mechanics. It was mentioned in Section 6.3 that some thought experiments can lead to apparent violations of the law. Some of these are related to electrodynamic systems; often in these systems momentum may appear to be created or destroyed, but what is overlooked is that the electromagnetic field can carry momentum, and while the solid particles in the system may see a net momentum change, the field itself changes to conserve the quantity.

In quantum mechanics, the notion of forces does not work but there is a quantum mechanical version of momentum, and this is conserved in all quantum mechanical processes. The conservation of momentum really is a law with no known exceptions!

It is useful to be able to not just understand and appreciate this proof but also to be able to reproduce it without assistance or reference to external sources. It is recommended that you take the time to learn how to reproduce the analysis leading from a statement of Newton's third law to the conservation of linear momentum before proceeding with the text. This is not an exercise in memory work (or rather, if it is, then that is the wrong way of going about it) but rather an exercise in developing logical reasoning to use one fundamental part of physics to prove another — something which all physicists must be able to do.

7.4. Using the Conservation of Linear Momentum

The law is extremely useful and important and you will use it throughout your career as a physicist. Just a few very simple

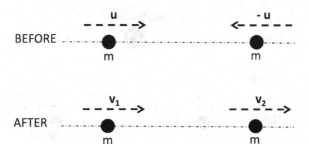

Figure 7.2:   A head on collision of two objects of the same mass. They masses have known speeds, u, in opposing directions, and unknown velocities, v_1 and v_2 after the collision.

examples will be given in this chapter but as the book goes on the law will be invoked with increasing levels of sophistication and complexity.

Example 7.1 Head on collision. If two objects of the same mass collide at the same speed but in opposite directions, they will either stop or rebound at the same speed but in opposite directions. A "before and after" diagram (Figure 7.2) can be used to analyse this.

The physics of the system is analysed as follows:

Total momentum of the system before the collision $= mu - mu = 0$

Total momentum of the system after the collision $= mv_1 + mv_2$

Hence, by using the conservation of linear momentum, $mv_1 + mv_2 = 0$.

The general solution to this equation is $v_1 = -v_2$ — i.e. the velocities of the two objects after the collision are equal and opposite. In the absence of any other information nothing more can be said about the velocities and they can take any value. The objects could simply stick together and stop for example, or bounce apart with the same speed they came in with, or any speed in between. They could rebound in a direction away from the original line if they are objects with an uneven shape but the vector nature of the solution indicates they remain along the same axis as each other.

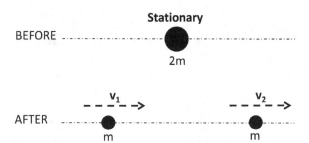

Figure 7.3:   **An object of mass $2m$ splitting into two pieces of equal mass.**

Example 7.2 Objects springing apart. If two objects that are initially stuck together spring apart due to an internal force they will move apart with equal and opposite momenta. If the objects have the same mass then the speeds will be the same. The before and after diagram is shown in Figure 7.3.

Analysis of the physics is very similar to the previous example, with the total momentum before the interaction being zero, and after being $mv_1 + mv_2$, so with the conservation of momentum this leads to $v_1 = -v_2$ — i.e. the velocities are equal and opposite again. As in Example 7.1, without any further information about the system the velocities can take any value.

An almost everyday example of such a system is two people on ice skates pushing themselves apart and moving away from each other at the same speed.

Example 7.3 Objects pulled together with a rope connecting them. If two people are on the ends of a rope and pull themselves together, the force is totally internal and the total momentum remains constant. The physics is as described in Section 6.2 on tension.

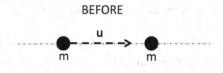

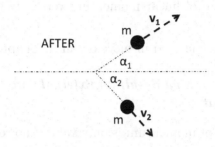

Figure 7.4: Two objects of the same mass in an angled collision.

7.5. Splitting Momentum Into Components

In $2D$ and $3D$ systems it is often convenient to split the momenta of the objects into their x, y and z components. Provided no external forces act in the direction of the component in question, linear momentum will be conserved along that component's axis.

For example, if a snooker ball (played without any spin) hits another of the same mass but the collision is not head on then the balls leave at a different angle to one other, such as in Figure 7.4.

The motion can now be split into horizontal and vertical components, with the horizontal ones being:

Total *horizontal* momentum of the system <u>before</u> collision $= mu$.

Total *horizontal* momentum of the system *after* collision $= mv_1 \cos \alpha_1 + mv_2 \cos \alpha_2$.

So applying the conservation of linear momentum horizontally gives $u = v_1 \cos \alpha_1 + v_2 \cos \alpha_2$.

The corresponding vertical components are:

Total *vertical* momentum of the system *before* collision is zero.

Total *vertical* momentum of the system *after* collision $= mv_1 \sin \alpha_1 - mv_2 \sin \alpha_2$.

The conservation of linear momentum vertically gives $v_1 \sin \alpha_1 = v_2 \sin \alpha_2$.

This problem is analysed in more detail in Chapter 10.

7.5.1. *Situations with a Resultant External Force Along One Component*

The conservation of momentum is only valid if no resultant external forces act on the system. But in a $2D$ or $3D$ problem, if such a force acts only along one component of the system then momentum is still conserved along the other "force free" components. This is of particular importance for systems under the influence of gravity near the surface of a planet with constant gravitational field strength — linear momentum will *not* be conserved perpendicular to the ground but will be conserved parallel to the ground.

7.6. **Two Classic Physics Puzzles**

There are a few well-known physics brainteasers that can be explained with reference to Newton's third law and the conservation of linear momentum. To round off this chapter, let us finish with two popular examples.

7.6.1. *The Sailing Boat and The Hair Dryer*

A person is stranded at sea on a calm day on a ship with a sail. They have a hair dryer with a long life battery that can provide a powerful blast of air. Explain why blowing the hair dryer at the sail will not help propel them towards the land. How *can* the hair dryer be used to manoeuvre the boat towards land?

First of all, let us consider what happens when there is a wind in the direction the ship needs to travel in. Let us say the ship needs to travel north. If there happened to be a wind blowing from south to north (a southerly wind) then the ship could be positioned such that that the wind hits the sail directly. When the wind strikes the sail, it stops — i.e. it changes momentum as the sail puts a force on the wind. By Newton's third law the wind therefore puts a force on the sail and it is this force that drives the ship forwards. The physics of this process is inspected in further detail in Section 10.7.

Now consider the hair dryer. In terms of classical mechanics, what a hair dryer does is take stationary air and push it out in a nearly uniform column at a certain speed. The hair dryer imparts a force on the air, and thus the air puts a force on the hair dryer in the opposite direction by Newton's third law. So with no sail present, to move north, the hair dryer simply needs to be pointed south and the force by the air on the dryer will propel the ship in the right direction.

But if the hair dryer is pointed north towards the sail two things happen: The force by the air on the dryer as it exits tends to push the ship south while the force by the air on the sail tends to push the ship north. In the event of no energy losses these forces will be equal and opposite in magnitude and the ship will go nowhere. In terms of the pure physics, the ship will be in state of tension while the hair dryer is on.

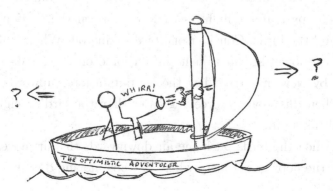

7.6.2.   *The Lorry Driver and the Geese*

A lorry driver with a load of geese in a sealed (yet somehow breathable) container comes to a weighbridge. The lorry is just over the weight limit, but without the geese the load would be ok. Explain why the driver's idea of creating a loud noise when on the weighbridge thus causing the geese to fly would not help in lowering the weight registered by the weighbridge.

Let us consider what the weighbridge actually records: It is the size of the contact force between the lorry's wheels and the bridge. It is possible to alter this contact force by, say, jumping. If instead of the flock of geese the driver had a single obedient human as cargo then if the human jumped at the right time, then during the jump the weight of the van would be increased during the push up part of the jump phase and lower when the human is in the air, as mentioned in Section 5.2 when describing the contact force of Newton's third law pair. This would not be practical of course — to beat the weighbridge, the person would have to be in the air precisely when the measurement occurred and weighing processes tend to take a few seconds. A human can only remain airborne during an unassisted jump for about a second.

Flying is different as the geese can remain airborne inside the truck for a comparatively long time. But it still would not work, as the key part about flight is that it relies on the motion of the air around the birds. Consider one bird taking off: When it flaps its wings in a downward motion it puts a force on the air downward, and so, by Newton's third law the air puts a force upwards on the bird. When this upward force is greater than the bird's weight then the bird takes off.

But when the birds push the air downwards the air moves down towards the floor of the lorry where it stops moving downwards and spreads sideways. The floor of the lorry thus puts an upward force on the air, and the air puts a downward force on the lorry.

What Newton's third law and the conservation of linear momentum tell us is that the reduction in downward force caused by the rising of the geese is exactly compensated for by the extra force by the down-moving air on the floor.

8

Work, Energy and Power

Having covered forces in detail, let us now move on to the equally important and familiar — though considerably more abstract — concepts related to energy. The use of work and energy in the study of physical systems allows a freedom of analysis that is often restricted by the use of forces alone. This chapter defines work and energy and shows how the work–energy theorem relates the work done on an object to its kinetic energy. It also defines mechanical power and derives a formula relating power supplied, force and velocity.

8.1. Work

8.1.1. *Definition, Units, and Values*

Consider an object under a constant force, F, that undergoes a displacement, r, *purely due to the action of that force*. In physics the amount of *work*, W, done *on* the object *by* the force is defined as the scalar product of the force with the displacement, i.e.

$$W = \boldsymbol{F} \cdot \boldsymbol{r} = Fr \cos \theta, \qquad (8.1)$$

where θ is the angle between the force and the displacement.

Work is a scalar quantity with SI units of joules (J) and base units $\mathrm{kgm^2s^{-2}}$. One joule is defined as the work done on an object when exactly $1\,\mathrm{N}$ of force acts on it over a straight line for exactly $1\,\mathrm{m}$.

All of the standard SI prefixes can commonly be seen in use with the joule.

If you lift a pen to write, you do about $10\,mJ$ of work on the pen; if a meteorite falls to Earth, the Earth does about $10\,GJ$ of work on the meteorite.

8.1.2.   *More on the Angle between the Force and the Displacement*

If the force is in the same direction as the displacement then the work done *on* an object *by* the agent causing the force is positive and given simply by Fr as $\theta = 0$ and $\cos 0 = 1$.

If the force is in the opposite direction to the displacement, then the work done *on* an object *by* the agent causing the force is negative and given simply by $-Fr$ as $\theta = \pi$ and $\cos \pi = -1$.

If the force is perpendicular to the displacement then the work done *on* an object *by* the agent causing the force is zero as $\theta = \frac{\pi}{2}$ and $\cos \frac{\pi}{2} = 0$.

For example consider an object being pulled straight along a table with a constant force, as in Figure 8.1.

In this simple example there are four forces on the block. The "pull" force is in same direction as the motion of the block, hence does positive work on it, while the friction force is in the opposite direction to the block and hence does negative work on it. As the contact force and weight are both perpendicular to the direction of motion they do exactly zero work on the block.

Another important example of a force being perpendicular to a displacement is if an object travels in a circle with an acceleration of constant magnitude — in this case, the infinitesimal displacements are always perpendicular to the force holding the object in a circle. This is examined in detail in Chapter 11.

If the force is at an angle θ to the displacement, then Equation 8.1 needs to be used in its general form. A good example would be if a block is pulled by a rope at an angle to direction of motion, as in Figure 8.2.

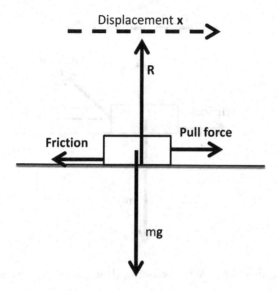

Figure 8.1: The pull force does positive work and the friction does negative work on the block. The contact force and the weight do zero work on the block.

In this case, the work done by friction, the contact force and the weight are as before but the work done by the pull force is now $\boldsymbol{P} \cdot \boldsymbol{r} = Pr\cos\theta$ from the defining equation.

Another way of thinking about it is to say the horizontal component of the pull force is $P\cos\theta$ and the product of this with the displacement gives the straight line work, whereas the vertical component of the pull force is $P\sin\theta$ and does no work on the block.

8.1.3. *Non-Constant Forces*

If the force acting on an object is not constant (either in direction or magnitude), then the small amount of work done, δW, when moving the object a small displacement $\delta\boldsymbol{r}$ is:

$$\delta W = \boldsymbol{F} \cdot \delta\boldsymbol{r}. \tag{8.2}$$

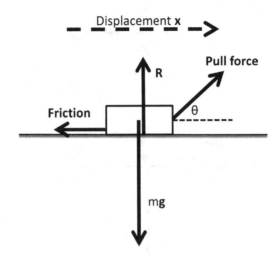

Figure 8.2:   A block being dragged at an angle.

So the total work done when moving from r_1 to r_2 is:

$$W = \int_{r_1}^{r_2} F(r) \cdot dr, \qquad (8.3)$$

where $F(r)$ is the force given as a function of position only. If we are lucky, the force will be integrable with respect to displacement and an exact solution is possible. In general, however, $F(r)$ will not be described by a mathematically expressible function.

As the force is an integral, it means that if the force acts in 1D only, the work done on an object by a force is given by the area under a force–displacement graph. For example, if the pull force in Figure 8.1 started at zero, rose to a maximum and fell again, the force–displacement graph might look something like that shown in Figure 8.3.

The area under the graph could be found by the "counting squares" method or some variety of computer approximation.

If the force were negative, then the curve would be "underneath" the displacement axis and the work done would be negative.

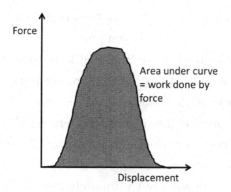

Figure 8.3: A force–displacement graph for a typical pulling force.

If the force is in 3D, then splitting the force into components and applying the technique to each component gives the work done.

A word of warning — it is easy to confuse force–displacement (area = work done) and force–time (area = impulse) plots. Compare Figures 8.3 and 6.1a. They could both represent the force imparted on a football when being kicked, for example, but show completely different things.

8.1.4. *Is the Work Done by Friction Positive or Negative? Some Words on Terrestrial Locomotion*

The answer to this question can be a difficult one to define as it depends on what the frictional force does. At first, it might be supposed that as friction always opposes motion it is therefore always negative. For example, if an object slides across a table until it stops then the work done by friction is definitely negative as the frictional force and the net displacement are in opposite directions. When sliding like this occurs, the work done by friction on the object is always negative, and energy is dissipated as heat. But friction does not always operate in this way.

Consider what happens when you run: You start from rest and then accelerate until you are moving at some velocity relative to the ground. In this case, the friction by the ground on you does *positive* work. When you move forwards you push your foot backwards on the ground. This push force is the frictional force *by* you *on* the ground. By Newton's third law, the ground therefore puts a frictional force *on* you in the forward direction and this is what drives you forwards.

This is the case for many forms of terrestrial motion, whether on feet, paws, hooves or wheels (see Chapter 17 for more on rolling motion) — in every case the energy is supplied by the animal or machine doing the moving but ultimately the external force causing the object to accelerate forwards is the frictional force by the ground on the moving object.

When slowing down, the friction then acts by doing negative work. When you are running fast and want to stop you jar your feet with the ground so there is a forward frictional force by you on the ground and thus a backward frictional force by the ground on you.

Finally, when walking at a uniform speed the net work done by friction must be zero: The push off with the back foot always sees a positive amount of work done by friction, and the landing front foot a corresponding amount of negative work.

8.2.   Energy, its Conservation, and Types of Energy

Although work has been rigorously and mathematically defined, it may not yet be clear what it actually means. Energy will now be introduced. The definition given may momentarily cloud the concept and not help with its abstract nature, nevertheless the more we learn about it, the clearer its importance will become. The definition that shall be used for this textbook is as follows:

> The energy of system is defined as its capacity to do work

Energy is a scalar quantity with the same units as work.

This statement means that if a system contains 1 J of energy then it has the capability of pushing a 1 N force in a straight line over a distance of exactly 1 m (or 0.1 N over 10 m or 10 N over 10 cm, and so on).

The importance of energy is that it is a *conserved quantity*. The **principle of the conservation of energy** states that **the total energy of an isolated system is constant.**

> **The energy of an isolated system cannot be increased nor decreased; it can only be changed from one form to another**

This principle is always true — there are no known exceptions. It will be used throughout this book, and indeed throughout your physics career.

There are several types of energy that you will have heard of (and no doubt know about in some detail). Of the main types light, sound, electrical, chemical, nuclear and heat energy fall under other subtopics within physics and will barely be mentioned here. In this book, mechanical energy is of the most importance, i.e. kinetic (this chapter), gravitational (Chapters 9 and 14), and elastic (Chapter 9).

8.3. Kinetic Energy and the Work–Energy Theorem

Kinetic energy is the energy due to motion. If one object is in motion relative to another one then it has the capacity to do work on the object (by colliding with it, basically. Whether the energy is usefully used depends on how controlled the collision is). This amount of energy is known as the moving object's kinetic energy and can be calculated as follows:

If a single force of any type is put on a free particle (i.e. no other forces on it), or the resultant external force on a particle is taken, then the force on it can simply be written as $= m\boldsymbol{a} = m\boldsymbol{v} \cdot \frac{dv}{dr}$.

The work, W, done on the object is given by Equation 8.3 — i.e. $W = \int_{r_1}^{r_2} m\boldsymbol{v} \cdot \frac{d\boldsymbol{v}}{d\boldsymbol{r}} \cdot d\boldsymbol{r} = m \int_{\boldsymbol{v_1}}^{\boldsymbol{v_2}} \boldsymbol{v} \cdot d\boldsymbol{v}$, where $\boldsymbol{v_2}$ and $\boldsymbol{v_1}$ are the final and initial velocities of the particle, respectively.

This is easily integrated to give the result that the work done on the particle is given by $\frac{1}{2}m\boldsymbol{v}_2^2 - \frac{1}{2}m\boldsymbol{v}_1^2$.

This change in energy is purely due to motion and is therefore the change in the kinetic energy of the object. If the particle starts from rest then the initial kinetic energy is zero and the absolute value for the kinetic energy, E_K, of an object relative to a stationary observer is:

$$E_K = \frac{1}{2}mv^2. \qquad (8.4)$$

En route to this we have also discovered another important rule:

> **The work done by the resultant external force on a particle equals the particle's change in kinetic energy**

This is the **work–energy theorem** and can be summarised mathematically by:

Work done by the resultant external force $= E_{K2} - E_{K1} = \Delta E_K.$
$$(8.5)$$

Notice that because the velocity is squared, the absolute kinetic energy of an object *must be greater than or equal to zero*. However, the *change* in an object's kinetic energy can be negative and in studying systems it is usually the change in quantities that are important. This is especially true with energy problems.

When physics students meet the work–energy theorem for the first time, it is common for them to consider it a special case of the conservation of energy. It is more than that. It is specifically to do with how the resultant external force on an object relates to an object's kinetic energy. It thus works alongside the conservation of energy as it brings in forces and dynamics — the conservation of energy on its own does not explicitly incorporate forces. The importance of the theorem is further developed in Chapter 9 when considering the vital distinction between conservative and non-conservative forces.

8.4. Power

A concept of great practical importance is the rate at which work is done. This is known as the **power**, P, generated by the force and according to the definition is given by:

$$P = \frac{dW}{dt}. \tag{8.6}$$

Power is a scalar quantity with units of Js^{-1} or watts (W) and base units of kgm^2s^{-3}. The power of a handheld electric fan would be a few watts, while the power of some modern day laser pulses in research laboratories are a few petawatts (though they cover a tiny area and last for a short time so the total energy provided is not great but concentrated into a small space and time).

All of the standard SI prefixes can commonly be seen in use with the watt.

Using the definition of work, an alternative formula for power can be derived as $P = \frac{dW}{dt} = \frac{d(\boldsymbol{F}\cdot\boldsymbol{r})}{dt}$ by the definition of work done.

The chain rule then gives $P = \boldsymbol{F} \cdot \frac{d\boldsymbol{r}}{dt} + \boldsymbol{r} \cdot \frac{d\boldsymbol{F}}{dt}$.

If the resultant external force is constant, then $\frac{d\boldsymbol{F}}{dt} = 0$ and $\frac{d\boldsymbol{r}}{dt} = \mathbf{v}$, by definition giving the result:

$$P = \boldsymbol{F} \cdot \boldsymbol{v}, \tag{8.7}$$

i.e. the instantaneous power supplied to a moving object is the dot product of the resultant external force with the object's velocity at that instant.

Power is extremely important especially when dealing with machines and other engineering concepts in applied subjects. It is of more limited use in developing a discussion of pure mechanical physics.

8.4.1. Does the Work Done When Lifting an Object Depend on How Fast it is Lifted?

The answer to this question is addressed properly in the next chapter, but before turning to it see if you can answer this yourself now.

9

Potential Energy

This chapter looks at the energy stored in a system that has the potential to do work. The general principle is introduced, and the two most important examples in mechanics — gravitational and elastic potential energy — are developed. There is also some discussion of force fields and the notion of conservative and non-conservative forces and their relation to potential energy. The chapter concludes with a discussion of the mass–energy equivalence relation.

Some of the analysis in this chapter is adapted from *Newtonian Mechanics* by A.P. French (1971).

9.1. Gravitational Potential Energy

Consider the motion of an object thrown upwards in a gravitational field. Once projected, the object is in free fall (i.e. the only force on it is the weight) and will move *upwards* and accelerate *downwards*. Let the object cross height y_1 at velocity v_1 and higher up cross height y_2 at velocity v_2, as illustrated in Figure 9.1.

The motion of the object can easily be described kinematically. Using Equation 3.12 gives $v_2^2 = v_1^2 + 2a(y_2 - y_1)$.

As the acceleration is simply $-g$, this becomes:

$$v_2^2 = v_1^2 - 2g(y_2 - y_1). \tag{9.1}$$

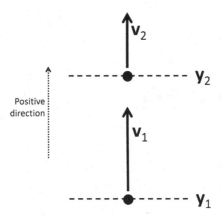

Figure 9.1: **An object in free fall moving upwards and accelerating downwards. In the diagram and the following analysis, positive is up and negative is down.**

In terms of work and energy, the change in kinetic energy between the two heights is given by Equation 8.4 as:

$$E_{K2} - E_{K1} = \frac{1}{2}mv_2^2 - \frac{1}{2}mv_1^2 = \frac{1}{2}m(v_2^2 - v_1^2). \qquad (9.2)$$

If Equation 9.2 is combined with the kinematic Equation 9.1, this gives $E_{K2} - E_{K1} = -mg(y_2 - y_1)$.

We know $-mg$ to be the gravitational force (the weight, W), hence the general result for the change in kinetic energy can be written:

$$E_{K2} - E_{K1} = W(y_2 - y_1).$$

This is in accordance with the work–energy theorem as this shows the change in kinetic energy is equal to the work done by the resultant external force on the object. In this case, the change in kinetic energy is negative as the object slows down; as the weight is negative and the change in displacement is positive, both sides of the equation are negative.

This can be rewritten so the two position 1 terms appear on one side of the equations and the two position 2 terms on the other:

$$E_{K2} + (-Wy_2) = E_{K1} + (-Wy_1).$$

This mathematical manipulation may seem trite and clumsy but has the effect of showing that the sum of two quantities is the same at the specified positions. As the positions have been selected arbitrarily it follows that the sum of the two quantities must be the same at *any* position the object passes through during its freefall.

This is a **statement of the conservation of mechanical energy**. If the quantity Wy is now *defined* as a new quantity which we shall call the **potential energy** (U) of the object such that

$$U(y) = -Wy, \tag{9.3}$$

it leads to the statement that the **total mechanical energy** (E_T) is given by:

$$E_T = E_{K2} + U_2 = E_{K1} + U_1. \tag{9.4}$$

An energy diagram for vertical motion above a surface will therefore be as shown in Figure 9.2.

As the weight is given by $-mg$, the potential energy can also be expressed by:

$$U = mgh, \tag{9.5}$$

where h is the height above or below a *reference point defined to be the zero of potential energy*. Note that other than the heading of the

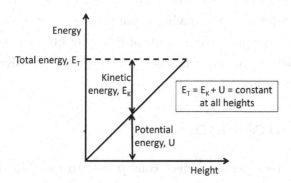

Figure 9.2: **Energies with height for vertical motion over a horizontal surface.**

subsection, the aim of the narrative was never to actually find an expression for potentially energy — it has simply appeared through careful analysis of the physics.

9.1.1.   *More Familiar Interpretation*

The potential energy is the energy due to an object's position in the gravitational field. It represents the amount of work that an external force (e.g. a contact force supplied by a person during lifting) would have to do in order to raise an object through a height h against the gravitational pull without providing it with any excess kinetic energy (i.e. the contact force supplied is only negligibly higher than the weight). If the object is then released, the gravitational force supplies the object with kinetic energy as it accelerates from $h \to 0$.

9.1.2.   *Potential Energy is Shared between Two or More Objects*

It is actually rather careless to refer to the potential energy of just one object in a gravitational field. In fact, the gravitational potential energy of an object on Earth is shared between the object and the Earth. It is incorrect to say that the potential energy belongs to either Earth or the object on its own. However, in everyday systems where small objects move about close to the surface of the Earth, it is fairly natural to refer to them having a potential energy of their own as the Earth barely moves as a result of the other object's motion. The shared gravitational potential energy between two massive objects is studied in Chapter 13.

9.2.   General Case in 1D

Now let us consider an object constrained to move in one dimension where the environment supplies a force that can be expressed as a function of displacement only such that $\boldsymbol{F} = \boldsymbol{F}(\boldsymbol{x})$. No more information is provided on the nature of the force. As the analysis follows it will be instructive to compare each step with the steps given in Section 9.1 to aid understanding.

If the object moves from a displacement x_1 to a displacement x_2 then the work done *by* the force *on* the object is given by $\int_{x_1}^{x_2} F(x) \cdot dx$.

By the work–energy theorem, this will be equal to the change in kinetic energy, $E_{K2} - E_{K1}$.

If an *arbitrary* reference point is now defined as x_0 then the rules of calculus allow us to write $\int_{x_1}^{x_2} F(x) \cdot dx = \int_{x_0}^{x_2} F(x) \cdot dx - \int_{x_0}^{x_1} F(x) \cdot dx$. It is simplest to imagine x_0 to be between x_1 and x_2 but it does not have to be in order to satisfy the mathematics.

Hence:

$$E_{K2} + \left[- \int_{x_0}^{x_2} F(x) \cdot dx \right] = E_{K1} + \left[- \int_{x_0}^{x_1} F(x) \cdot dx \right].$$

This is the generalised version of the "trite and clumsy" step from the discussion in Section 9.1.

The potential energy, $U(x)$, for this general force can then be *defined* by:

$$U(x) - U(x_0) = - \int_{x_0}^{x} F(x) \cdot dx, \qquad (9.6)$$

which is the general version of the step to obtaining Equation 9.3.

Verbally, this equation means that the potential energy at a point, relative to the reference point is defined as the *negative* of the work done *by* the force *on* the object as the object moves from the reference point to the new point x.

An equivalent way of writing Equation 9.6 is in the differential format:

$$F(x) = -\frac{dU}{dx}, \qquad (9.7)$$

i.e. the force is the negative gradient of the potential energy — a very powerful formula!

When starting to study mechanics, it is sensible to study forces and use knowledge of a system's dynamics to obtain information about the energy. The deeper we go into the subject, however, the more the energy takes the role as the fundamental quantity and the force becomes (literally) the derived quantity. In quantum mechanics for example the notion of forces is not really very sensible, however the motion of objects as a result of the potential energy gradients is still valid. Equation 9.7 and its more general variant

$$F = -\nabla U, \qquad \textbf{(3D version of Equation 9.7)}$$

(the 3D version which uses vector calculus — we will not use this version which has merely been included for completeness) are the most important equations in this book when put in the context of the whole of physics. Within the confines of the pages of this book $F = ma$ is the most important equation, but within the whole of physics $F = -\nabla U$ assumes the more fundamental status and the greater importance.

9.3. Elastic Potential Energy

This section is essentially an example that shows how the ideas developed in Section 9.2 can be used to derive the potential energy of a common and familiar mechanical system: The stretched spring. In fact, the use of Equation 9.7 to derive the formula for elastic potential is like using a sledgehammer to crack a nut, but it will do nicely to illustrate how it can be used.

Consider a spring with a *restoring force* that is directly proportional to its displacement. The force acts in the opposite direction to the displacement, hence the force is given by:

$$F = -kx, \qquad (9.8)$$

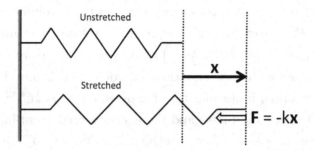

Figure 9.3: Stretching a spring from equilibrium.

where k is a scalar constant in Nm^{-1} usually known as the *spring constant*. Its value depends on the material and dimensions of the spring.

Consider that one end of the spring is attached to a fixed point and points along the $x-$axis, and is then stretched in the direction of positive x as shown in Figure 9.3.

Then if the unstretched spring is defined to have zero potential energy then the potential energy of the stretched spring will, from Equation 9.6, be:

$$U(x) = -\int_0^x -kx' \cdot dx' = k\int_0^x x' \cdot dx'$$

and so:

$$U(x) = \frac{1}{2}kx^2. \tag{9.9}$$

A familiar result!

9.3.1. *Stored Energy* $= \frac{1}{2} \times$ *Constant* $\times$ *Variable*^{2} *Formulae Appear Quite a Lot in Physics*

Linear kinetic energy is given by $\frac{1}{2}mv^2$. This is essentially the energy stored by a moving object. We will see in Chapter 13 that rotational kinetic energy — a measure of the kinetic energy stored in a spinning object — is given by $\frac{1}{2}I\omega^2$ where I is a constant and ω is the object's angular velocity. The energy stored in a spring is $\frac{1}{2}kx^2$.

In electromagnetism, the energy stored per unit volume in an electric field, E, is given by $\frac{1}{2}\epsilon E^2$ and the energy stored per unit volume in a magnetic field, B, is $\frac{1}{2}\left(\frac{1}{\mu}\right)B^2$ where ϵ and μ are constants. These are probably new equations to you but you may have seen the energy stored in a capacitor of capacitance C is $\frac{1}{2}CV^2$ where V is the PD across the plates and the energy stored in an inductor of inductance, L, is $\frac{1}{2}LI^2$ where I is the current flowing. These are both specific versions of the electric and magnetic field equations.

The stored energy $= \frac{1}{2} \times$ constant $\times$ variable^2 formula is therefore quite a common occurrence across the discipline. It tends to arise when a force or force-type quantity increases linearly with displacement (as with the spring). To get the stored energy, an integral with respect to the distance then raises the power of the distance from $1 \rightarrow 2$ and puts a factor of $\frac{1}{2}$ in front.

9.4.   Conservative and Non-Conservative Forces

9.4.1.   *Introduction*

Only certain forces have associated potential energies. As well as the gravitational and spring forces, other "action-at-a-distance" forces (e.g. electrostatic) will have potential energies. Forces of this kind are classified as **conservative forces**.

If a force does not have an associated potential it is a **non-conservative force** and this accounts for all the macroscopic contact-type forces we have seen in classical mechanics.

For a force to be conservative both the force and the associated potential energy must be well-defined, single-valued functions of position. Consider frictional forces, for example, the magnitude of a frictional force on an object certainly does not solely depend on an object's position, therefore the frictional force is non-conservative.

9.4.2. *Other Properties*

Conservative forces have associated fields, and an object in the field will experience a force defined solely by its position. It can be proved that the work done by a conservative force in moving from one point to another in the field is *path independent* (i.e. if you move from point A to point B, the amount of work you do does not vary with the route you take) and a corollary of this is that the work done moving in a closed loop is zero (i.e. if you move from point A, travel around and return to point A, the net work done is zero regardless of the route).[1] This is probably best pictured by thinking of an object moving in the vicinity of the gravitational field of a planet. The same cannot be said of non-conservative forces.

If a non-conservative force acts on an object then the object's mechanical energy ($KE + PE$, Equation 9.4) will change; if the work is positive it will gain mechanical energy, if negative it will lose it. However, if a conservative force acts on an object, the total mechanical energy will always remain constant.

9.4.3. *Lifting a Box*

At the end of the previous chapter you were asked to consider the energy changes involved in lifting an object. We are now in a position to address the answer properly.

Consider lifting a box so that it starts and finishes with zero kinetic energy. This could be done by lifting slowly and carefully with the box rising at near-negligible speed, or the normal way by lifting relatively quickly at first and slowing to zero speed at the desired height. The change in the object's *kinetic energy* equals the *total* work done by *all* the forces (weight and contact force) and is

---

[1]The mathematics required for a proper analysis is part of the subject known as vector calculus and is covered in all physics degree courses, usually in the first year.

zero: the contact force does positive work and the gravitational force does negative work.

But the change in the object's *total* energy (kinetic plus potential) equals the work done by the *non-conservative forces*, in this case the contact force. This is positive, so the object's total mechanical energy increases, in this case solely by an increase in potential energy.

At GCSE level, if you are asked about the work done in lifting a box you would probably have considered it a terribly simple problem, but now you realise there is a lot more to it than a mere Work done = PE gained = mgh type of answer.

9.5.   Potential Wells

If an object is in a conservative force field then it will always accelerate towards an equilibrium point where the net force on it is zero. This will always be a local stationary point of potential energy (from Equation 9.7, $F = 0$ when $\frac{dU}{dx} = 0$ so it needs to be a maxima or minima on a U vs. x plot), and will only be stable at this point if the potential is a minimum.

This can be simply seen by plotting some conservative forces and their corresponding potentials shown in the sketch plots in Figures 9.4 to 9.7.

1. Lifting a weight

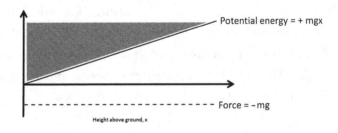

Figure 9.4:   Potential well for lifting a weight.

2. Stretching a spring

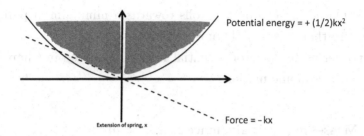

Figure 9.5: Potential well for stretching a spring.

3. Object attracted to the Sun

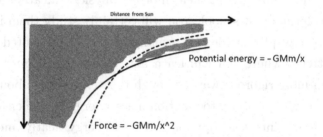

Figure 9.6: Potential well for gravitation of a point mass.

4. Atom–atom interaction

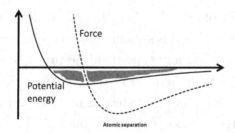

Figure 9.7: Potential well for two atoms (known as the Lennard–Jones potential).

The shaded area on the figures is known as the *potential well.* The motion of a particle in the potential can be thought of as a marble that is placed on a slope and rolls towards a minimum of potential (take care that this is a 1D motion though).

Another useful facet of potential wells will be seen in Chapter 11 on simple harmonic motion.

9.6.   Mass–Energy Equivalence and $E = mc^2$

Most readers will have heard of mass–energy equivalence and Einstein's famous equation $E = mc^2$ (where E is loosely defined as simply "energy", m as mass and c is the speed of light and is equal to $3 \times 10^8 \, \text{ms}^{-1}$) as part of their informal physics education and possibly also formally at school. You will certainly study them in detail as part of any physics degree, usually in a course devoted to special relativity, where you will learn how to prove $E = mc^2$ both in quick and more rigorous ways. Though this textbook is about classical mechanics, it is useful to mention mass–energy equivalence at this stage as it fits into the topic of potential energy neatly and this is also a useful point at which to clear up some common misconceptions about the way it works.

9.6.1.   *Mass–Energy in General*

Consider a system of particles of any kind. Let the mass of the system of particles be m. Now if this system of particles is given an additional amount of energy E of any type then the principle of mass–energy equivalence states that the mass of the system of particles increases by $\frac{E}{c^2}$ so the new mass of the system of particles is $\left(m + \frac{E}{c^2}\right)$.

As c^2 is so large, for most everyday situations the mass changes are negligible, as shall be seen in the examples that follow. However, this law has been verified time and again by experiment since Einstein's famous paper on the subject in 1905 and as far as we know there are no known exceptions to it.

The law is completely general — it applies to any system of particle and for any type of energy addition. To clarify, here are some everyday examples:

9.6.2. *Stretching a Spring*

Imagine that there is an unstretched spring inside a box. The box has hooks at each end. The box, the spring and the hooks comprise the "system of particles" referred to in the general case. A person now measures the mass of the system. It is not important how this is done — it could be by placing it on a balance, or by pushing it with a known force, measuring the acceleration and finding the mass by $m = \frac{F}{a}$. Whatever the method, imagine the mass can be measured to an arbitrarily high accuracy and precision.

An external agent (e.g. a person) now stretches the spring, attaching each end to the hooks at the ends of the box so it is in a permanent state of extension. The person has done work on the system — i.e. it has been given an amount of energy. In this case, the energy can be described as elastic potential energy, which is ultimately electro–magnetic in nature and can be quantified as being of a value $\frac{1}{2}kx^2$ with the symbols as in Equation 9.9. As energy has been added, the mass has now increased — by an amount $\frac{\frac{1}{2}kx^2}{c^2}$ — and if the mass is measured by whatever means, it will be observed to have increased by this amount.

Of course, we do not notice that springs get heavier when we stretch them as the amount of mass increase is incredibly small. For a spring constant of $1\,\mathrm{Nm}^{-1}$ and a stretch of $10\,\mathrm{cm}$, the increase is a mere $10^{-19}\,\mathrm{kg}$. One would need an accurate experiment indeed to detect the change.

9.6.3.   *Charging a Battery*

Consider an uncharged rechargeable battery as the system of particles with the mass measured with the same accuracy and precision
as before. The battery is then "charged up". The mechanism for
this process is not important for this discourse, but in GCSE-level
terminology, we can say the battery has now gained chemical potential energy. If the battery is now removed from the charger and
the mass taken again it will be higher. Once again, the increase is
small: The amount of energy that a small battery is capable of stor-
ing is of the order of 10^6 J, so the mass increase is of the order of
$\frac{E}{c^2} = \frac{10^6}{(3 \times 10^8)^2} \cong 10^{-11}$ kg. As batteries have a mass of a few tens of
grams, this is a tiny effect.

9.6.4.   *Kinetic Energy, Dissipation of Heat, and Cups of Tea*

The law also works for kinetic energy. Consider the system of particles
to be a snooker table with the snooker balls placed on the surface
prior to taking a shot. The mass is now taken and recorded with the
same arbitrarily good accuracy and precision. Now an external agent
(a snooker player) plays a shot which imparts kinetic energy to one or
more of the snooker balls in the system. If the mass is now recorded
while the balls are in motion it will be higher than before. You can
compute for yourself what a small effect this will be. Eventually of
course the balls come to rest and the mass will eventually be the same
again. The kinetic energy of the balls ultimately gets dissipated as
heat and "sound" energy.

The sound energy is essentially an increased kinetic energy of the
air molecules in the vicinity of the balls which radiates outwards — if
the experiment were performed in a vacuum (an admittedly unlikely
scenario) then the collisions would not involve any noise and the
snooker balls would dissipate their energy, essentially by friction and
thus as heat, by collisions with each other and the walls of the table.

It is worth discussing what happens when energy is dissipated as heat. The heating effect occurs because the vibrational kinetic energy of the atoms making up the balls increases and this gradually reduces, partly by transferring the kinetic energy to surrounding air molecules, and partly (wholly if in a vacuum) by gradual emission of low frequency electromagnetic radiation. We know it must be low frequency as the snooker balls do not glow after even the most powerful shot, so the emission of radiation is below the visible range.

A similar thing happens when a cup of tea cools. If the cup has a lid so no water can escape by evaporation, the heat energy (manifest as vibrational, translational and rotational kinetic energy of the atoms and molecules making the tea and the mug itself) gradually reduces as kinetic energy is passed on to the surrounding air molecules and radiated outwards as electromagnetic radiation. This still is not visible but does reach infrared frequencies — there are free smartphone apps which allow you to see this.

So a hot cup of tea weighs more than a cold one provided no liquid is lost by evaporation. However, if the cup of tea were kept in a thermally insulated black box with a vacuum inside there would be no change. In such a situation, the box would not release any of the radiation from the tea, which would gradually accumulate electromagnetic radiation (i.e. the box would be "full of photons"). Unless this radiation is released all the mass will still be contained within.

9.6.5. *Climbing a Mountain*

One of the sloppiest examples of the use of $E = mc^2$ is in relativity questions that state something along the lines of "A 60 kg person climbs a mountain of height 2 km. Calculate their increase in mass and comment on your answer".

The purpose of this type of question is to highlight how small a gain in mass the person undergoes — the gravitational potential energy gained is $mgh \cong 1.4\,MJ$ so the mass "gained" is $\frac{1.4 \times 10^6}{(3 \times 10^8)^2}$ $\approx 10^{-10}$ kg, which is significantly smaller than a hair on your head, therefore the mass gained is negligible and so on.

Unfortunately, this kind of question can be seen in some physics textbooks right up to undergraduate level and can still be posed in physics classrooms from time to time.

The main reason this question is so bad is that it does not properly consider the person and the Earth as a single system of particles, which is how it must be in order to properly invoke $E = mc^2$. The person's mass does not really change at all, but the mass of the person–Earth system *might* change, depending on what is measured.

An explanation of how the physics could be discussed properly in two different ways is as follows:

Case 1: The person receives a lift from an external agent

Consider a person standing at the bottom of the mountain, and consider the system of particles to consist of just two objects: The person (of mass m) and the Earth (of mass M). Any external observer would measure the mass of the person–Earth system to be $(M + m)$.

Now, an alien spacecraft comes to Earth, picks up the person and does work on the person–Earth system to pull the person away from the Earth and place him or her on top of the mountain. If the height of the mountain is H, the spacecraft does an amount of work mgH on the system. Note that it would be incorrect to say that the person now has that much potential energy — it is shared between the person and the Earth. It is also incorrect to say that the person's mass has increased — he or she is still made of the exact same protons, neutrons and electrons as before. The Earth's mass is the same too.

But the outside observer with the extremely sensitive mass-measuring equipment *does* measure a difference — the mass of the whole system is now recorded as $\left(M + m + \frac{mgH}{c^2}\right)$.

Case 2: The person climbs the mountain

This time the external observer sees the mass as $(M+m)$ all the time. Let us imagine the person climbs completely efficiently — i.e. every joule of energy used by the person is converted to 1 J of gravitational potential energy. In this case, the person starts out with an amount of stored chemical–potential energy in his or her body. As they climb, this is converted into gravitational potential energy shared between the person–Earth system. When the person reaches the top they have lost mgH of stored chemical potential energy, which has been gained by the system. The total mass of the system remains the same.

In this case, the person has lost a significant amount of mass by respiration (which unless it has escaped the Earth's system will still be recorded by the external observer) and an insignificant amount from the loss of stored energy $\frac{mgH}{c^2}$.

For the real world case of the person radiating heat as they climb, unless the radiation escapes the Earth this will also be measured as the mass of the system by the external observer.

Hopefully, it should be realised now that the notion that the person gains mass due to their ascent is completely misleading, and in fact the opposite is true if the person ascends the mountain without external assistance.

9.6.6. *Combustion, Breathing, and Weight Loss*

In chemical reactions, bonds between atoms and molecules are broken and reformed during a process starting with one set of compounds and ending with another. The bonds can be classified in various ways (as you may well know from chemistry lessons, with covalent, ionic and metallic being the most discussed at school level) and all differ in detail but have one overarching similarity in that they are all essentially electromagnetic in nature.

In endothermic reactions, some energy needs to be added to the start products for the reaction to occur and in exothermic reactions some energy is liberated during the process. Let us consider how $E = mc^2$ for combustion, i.e. a simple exothermic process.

Consider the combustion of the simple hydrocarbon methane. Methane combines with oxygen to produce water and carbon dioxide. The chemical reaction for the process is:

$$CH_4 + 2O_2 \rightarrow 2H_2O + CO_2 + \text{energy}.$$

In GCSE-level terms, the energy liberated is in the form of heat and light (i.e. fire) and in slightly more advanced language you could say it is a range of frequencies of electromagnetic radiation. The exact nature of the mechanism for the process is not important for the analysis — we simply need to recognise that energy is released.

It is found experimentally that approximately $810\,kJ$ of energy are released in burning 1 mole of methane. This means that $\frac{E}{c^2} = \frac{810 \times 10^3}{(3 \times 10^8)^2} = 9 \times 10^{-12}\,kg$ of mass is "lost" energy when that amount of methane burns. So if the mass of the initial methane and oxygen was accurately taken and compared with the final mass of water and carbon dioxide, the mass deficit could, in principal, be noted. However, this would only be if the energy escaped the system. If it could somehow be contained (by having the combustion take place inside a thermally sealed container) then the mass of the whole system would not be observed to change.

Note that in the chemical reaction the number of elementary particles is no different before and after the reaction. The "amount of stuff" is the same, meaning the early definition of mass is not valid for mass–energy relationships. No particle mass has actually been lost at all in the process — all that has happened is a liberation of potential energy.

It is commonly said that when one exercises one burns up excess fat and this causes us to lose weight. While this is true in some sense, the order of magnitude estimation for the mass loss given above should make it clear that weight loss cannot happen simply by radiating out the energy. When we breathe, we also burn hydrocarbons. The process is much more complicated than the combustion of methane outlined above but if a basic respiration equation using glucose as the start point is:

$$C_6H_{12}O_6 + 6O_2 \rightarrow 6CO_2 + H_2O + \text{energy}.$$

The energy is used for doing work, but that is not where the weight loss part comes in — rather it is the exhalation of carbon dioxide that reduces the weight. For every oxygen molecule we breathe in, we breathe out two oxygen atoms attached to a carbon atom and this is where the weight loss occurs — the exhalation of carbon atoms. Essentially, to lose weight you need to breathe more!

9.6.7. *Nuclear Reactions*

In the nucleus of an atom the protons and neutrons are bound together by the *strong nuclear force*. This is essentially what prevents the protons from flying apart by electrostatic repulsion. The *nuclear binding energy* of the nucleus is the mechanical energy needed to separate all of the particles making up the nucleus so they are a long distance apart. "Long distance" really means an infinite distance, but in reality would mean a distance great enough so the strong nuclear force becomes negligible. The binding energy is thus a positive number of joules; it is equal in magnitude to the energy liberated if the constituent particles come together from a long distance apart to form a nucleus.

In a nuclear reaction, particles are prised apart (requiring an input of energy) and then reassembled (resulting in a release in energy). If the net energy change is positive, then the overall mass of the system of particles reduces according to $E = mc^2$. In many nuclear reactions, the mass change is a significant proportion of the original mass of the system.

Another unfortunate misconception regarding mass–energy equivalence is that in nuclear reactions mass is simply turned into energy. A common thing to do at school is compute that just 1 g of mass can liberate a staggering $mc^2 = 10^{-3} \times (3 \times 10^8)^2 \approx 10^{14}$ J of energy, so if we could properly harness this, we could solve the world's energy crisis. The issue is that there are not many reactions that simply convert mass to energy, and certainly not any that we can currently harness on a large scale. When matter meets antimatter (a proton interacting with an antiproton for example) then this can result in complete annihilation (that is the technical term) of the particles, producing energy in the form of electromagnetic radiation (photons). Though antimatter particles are created and subsequently annihilated in some nuclear reactions, the majority of man-made nuclear reactions have the same number of particles before and after the reaction — the energy is liberated through a reshuffling of subatomic particles to reduce the potential energy of the system and radiate some photons in the process. The amount of energy liberated can never match the huge numbers promised by simple annihilation processes.

10

Collisions and Rockets

This chapter applies the physics developed in the preceding chapters to several situations involving energy and linear momentum transfer. This is the first chapter when no new fundamental principles will be introduced, but several important results are elucidated. For each situation, the general principle will be introduced and the occasional example given.

This is one of the longest chapters in the book as we extract plenty of physics out of simple situations using the tools we now have at our disposal.

10.1. Collisions

Let us first consider collisions between two particles. For simplicity's sake, in these collisions the particles are not rotating and remain intact during the collision with no transfer of mass from one particle to the other. In physics, in a collision between two particles:

(1) The particles exert a force on each other according to Newton's third law.
(2) The Newton's third law pair of forces exists for a finite amount of time — i.e. the collision has a beginning and an end.
(3) The particles do not have to touch, but may do so, at any point in the collision.

The most convenient collision to envisage is a two dimensional collision between two spheres (like snooker balls, for example) and

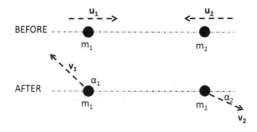

Figure 10.1: **Before and after diagram for a generic two particle collision.**

most of the examples can be thought of this way. In such a collision, the force will have the general form illustrated in Figures 7.1a and 7.1b for an impact between two objects.

Consider, for example, two particles of masses m_1 and m_2 that collide with initial velocities u_1 and u_2 and have velocities v_1 and v_2 after the collision, as in Figure 10.1.

It is usual, though not essential, to choose the axes so that the particles approach each other along the same axis. It is also possible to choose the reference frame so that one of the particles is stationary (refer to Section 10.2). Note that it is always possible to do this. Think of two snooker balls travelling towards each other and colliding. You can always choose your reference frame so that you travel alongside one of the balls before it collides. This ball will be stationary before collision and the other will do the moving. After the collision you continue at the same velocity and, in general, both balls will move away.

Subsections 10.1.1–10.1.3 apply the above general picture to three specific cases.

10.1.1. *Elastic Collisions*

Some collisions between particles are referred to as elastic. Sometimes clarifying adjectives are added so they can be referred to as perfectly elastic, completely elastic or totally elastic. All mean the same thing.

By definition, in a perfectly elastic collision:

(1) Total energy is conserved.
(2) Kinetic energy is conserved.
(3) Linear momentum is conserved.

Considering the general case illustrated in Figure 10.1, using the conservation of linear momentum with the formula $p = mv$ for momentum:

$$m_1\boldsymbol{u_1} + m_2\boldsymbol{u_2} = m_1\boldsymbol{v_1} + m_2\boldsymbol{v_2}.$$

Using the **conservation of energy and the fact the collision is elastic** so kinetic energy is conserved and using the formula $E_K = \frac{1}{2}mv^2$ gives $\frac{1}{2}m_1u_1^2 + \frac{1}{2}m_2u_2^2 = \frac{1}{2}m_1v_1^2 + \frac{1}{2}m_2v_2^2$, which multiplies through by 2 to give:

$$m_1u_1^2 + m_2u_2^2 = m_1v_1^2 + m_2v_2^2.$$

These boxed equations are completely general and apply in three dimensions — i.e. if two particles ever collide elastically, these equations hold. For simplicity, two dimensions will be used in the following analysis.

10.1.1.1. *2D case with components*

Often it is convenient to split the momentum into perpendicular components, giving:

(1) $m_1u_1 + m_2u_2 = m_1v_1 \cos\alpha_1 + m_2v_2 \cos\alpha_2$ for the horizontal components,
(2) $m_1v_1 \sin\alpha_1 = m_2v_2 \sin\alpha_2$ for the vertical components,
(3) $m_1u_1^2 + m_2u_2^2 = m_1v_1^2 + m_2v_2^2$ for the energy as before.

It *is* possible to solve these equations generally — there are three equations and three unknowns.[1] However, the algebra gets rather

---

[1] The three unknowns are the two velocities after the collision and the angle of separation after collision. The way the problem is set up, it appears that there are two unknown angles, but knowing one always defines the other. Example 10.2. shows a specific example of this.

messy and the resulting formulae are cumbersome and not very useful.[2] It is far better to stick with the starting principles and apply these to specific situations.

Example 10.1 Results for a 1D elastic collision — i.e. the particles collide along a straight line. A particle of mass m_1 travelling at velocity v collides elastically and in a straight line with a stationary particle of mass m_2. Find the velocities of both particles after the collision.

Free body diagrams are not particularly useful with problems such as this as the nature of the forces is not important to the problem. However, before and after diagrams showing the particles' velocities (Figure 10.2) are very handy.

Taking the inertial reference frame as given in the image: Using the conservation of linear momentum, total momentum before the collision = total momentum after the collision, so:

$$m_1 u = m_1 v + m_2 w. \tag{10.1}$$

As the collision is stated to be perfectly elastic, kinetic energy is conserved so $\frac{1}{2} m_1 u^2 = \frac{1}{2} m_1 v^2 + \frac{1}{2} m_2 w^2$, giving:

$$m_1 u^2 = m_1 v^2 + m_2 w^2. \tag{10.2}$$

Figure 10.2:   Before and after diagram for a 1D elastic collision.

---

[2]This is not a hard fact and is more the author's opinion. Try it if you wish! You will need a lot of paper and some patience and ultimately you may disagree and find it a rewarding exercise. The author would be pleased to hear any feedback on this.

These are two equations with two unknowns which can be solved[3] to give:

$$v = \frac{m_1 - m_2}{m_1 + m_2} u,$$

$$w = 2\frac{m_1}{m_1 + m_2} u.$$

These lead to some nice results for special cases. You can test all of these in a very rough way with a flat, smooth tabletop and some coins by flicking one coin into another. For a large mass and a small mass, a penny piece and a pound coin works well, and any two coins of the same size work well for testing two similar masses.

(1) If $m_1 \gg m_2$ then $\boldsymbol{v} \approx \boldsymbol{u}$ (but a bit smaller) and $\boldsymbol{w} \approx 2\boldsymbol{u}$ — i.e. the big mass barely slows down but the small mass pings off with approximately *double* the speed of the big mass. This means that no matter how large you make the big mass, there is an upper limit to the speed with which the small mass moves away.

   Try this with the coins — pound striking penny. It can be a bit difficult to see that it is double the speed (or nearly double), but you should just about be convinced!

   This result can be intuited if one thinks about moving along with the reference frame of the big mass. From the big mass's point of view, it is stationary, and the small mass approaches *it* at a speed, v. What the big mass experiences is merely that the small mass rebounds at nearly the same speed while the big mass moves back at a tiny velocity. So the big mass sees the small mass keep its speed and reverse its direction, which is the equivalent of observing the small mass go from stationary to twice the speed of the big mass in the original reference frame.

---

[3]This is an exercise for the reader which is much easier and requires less paper and time than the previous footnote. If you are unsure, start by making w the subject of the formula in Equation 10.1. Then square both sides, substitute the resulting expression for w^2 into Equation 10.2 and rearrange.

(2) If $m_1 = m_2$ then $v = 0$ and $w = u$ — i.e. the velocities are transferred.

Try this with two identical coins — provided they strike in a straight line the result will be shown nicely.

(3) If $m_1 \ll m_2$ then $v \approx -u$ (but a bit smaller in magnitude) and $w \ll u$, but is small and positive — i.e. the small mass rebounds with the same speed and the big mass hardly moves, which is intuitive.

This is simple to show with a penny striking a pound, and just involves the penny rebounding. This is very intuitive and is, in a sense, the opposite of the first situation.

Example 10.2. The 90° rule. In a game of marbles, a marble rolls across a table and strikes another stationary marble of the same mass. A negligible amount of energy is dissipated during the collision. Show that whatever the angle of impact, the marbles will move apart perpendicular to each other.

The before and after diagram for the marbles is shown in Figure 10.3.

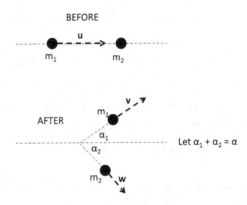

Figure 10.3:   Before and after diagram for a 2D elastic collision.

Using the conservation of linear momentum, momentum before = momentum after so using $p = mv, mu = mv + mw$, hence:

$$u = v + w. \tag{10.3}$$

Note that the vector notation is deliberately retained in this equation.

As negligible energy is lost, the collision is completely elastic and kinetic energy is conserved, hence using $E_K = \frac{1}{2}mv^2, \frac{1}{2}mu^2 = \frac{1}{2}mv^2 + \frac{1}{2}mw^2$, hence:

$$u^2 = v^2 + w^2. \tag{10.4}$$

There are two equations with two unknowns.

Squaring Equation 10.3 gives: $u^2 = (v + w)^2 = v^2 + 2v \cdot w + w^2$
$$= v^2 + w^2 + 2vw \cos \alpha.$$

Substituting this expression for u^2 into Equation 10.4 gives:

$$v^2 + w^2 = v^2 + w^2 + 2vw \cos \alpha,$$

which implies $\cos \alpha = 0$, hence $\alpha = \frac{\pi}{2}$ — i.e. the velocities after the collision are perpendicular.

This is easy to verify in practice, at least to a 5% level of accuracy. Try it with two identical coins, drawing lines to show the route they take after the collision, and look at the angle between them — it will be approximately 90° (probably a little less). This works no matter what collision angle you take.

The problem could also be solved using components instead of vectors, but is less elegant and involves more clumsy algebra.

10.1.1.2. *Instances of elastic collisions*

Marbles, snooker balls and coins colliding are not really perfectly elastic collisions.[4] Collisions will be completely elastic if the forces

---

[4]The author's personal tutor at the University of Bristol, Dr. Derek Parsons, used to summarise the argument that they were not with one word: "Click!"

between the objects that collide are conservative, and in practice this means "action at a distance" forces only. If objects actually connect during a collision, then there are non-conservative contact forces and frictional forces acting as Newton's third law pairs between the objects, which inevitably leads to loss of kinetic energy as a consequence of the work–energy theorem. Typically, these losses are as heat energy, sound energy and energy lost through permanent elastic deformation on one or both of the colliding objects (think of two cars crashing into each other with the bodywork getting bent out of shape).

So when are collisions perfectly elastic? On a human size scale, almost never, but there are very large scale and very small scale examples. On a large scale, the "collision" between a comet and a star would be perfectly elastic. The force between them is gravitational and they never contact. There is no real beginning and end to this collision, but there will be a distance when the attractive forces can be considered negligible and elastic collision theory can happily be applied. On a small scale, meanwhile, collisions between charged particles are perfectly elastic as the particles do not actually touch. Astrophysicists and particle physicists regularly deal with perfectly elastic collision without any worry of experimental uncertainty. On a human scale it is good for a first approximation but normally we need to be aware that there is some energy lost that will affect the result.

10.1.2. *Inelastic Collisions*

By definition, in an inelastic collision:

(1) Total energy is conserved.
(2) Kinetic energy is lost.
(3) Linear momentum is conserved.

Thus, for a two-body collision the linear momentum expression remains the same as the general case in Subsection 10.1.1, but the kinetic energy expression is modified to:

$$\frac{1}{2}m_1 u_1^2 + \frac{1}{2}m_2 u_2^2 = \frac{1}{2}m_1 v_1^2 + \frac{1}{2}m_2 v_2^2 + \epsilon,$$

where ϵ is the positive kinetic energy dissipated as a result of the collision.

If two particles collide and coalesce (i.e. stick together) the collision is said to be *completely inelastic* (also totally inelastic or perfectly inelastic).

Example 10.3. The completely inelastic head-on collision. A particle of mass m_1 travelling at velocity u collides with a stationary particle of mass m_2. Find the maximum possible energy loss in the collision and show that in this case the particles stick together.

The before and after diagram for the particles is shown in Figure 10.4.

Using the conservation of linear momentum, momentum before = momentum after so using $p = mv$:

$$m_1 u = m_1 v + m_2 w. \qquad (10.5)$$

Let the amount of energy lost in the collision be ϵ. Using the conservation of energy and $E_K = \frac{1}{2}mv^2$, $\frac{1}{2}m_1 u^2 = \frac{1}{2}m_1 v^2 + \frac{1}{2}m_2 w^2 + \epsilon$, hence:

$$m_1 u^2 = m_1 v^2 + m_2 w^2 + 2\epsilon. \qquad (10.6)$$

There are two equations with two unknowns.

BEFORE   AFTER

Figure 10.4: Before and after diagram for a 1D inelastic collision.

The algebra is awkward — try it if you like, but only if you feel you need the practice! — by using Equation 10.5 to find an expression for v^2, then substituting into Equation 10.6 renders a quadratic equation for w. The positive solution of this quadratic is:

$$w = \frac{m_1 m_2 u + \sqrt{m_1^2 m_2^2 u^2 - 2\epsilon m_1 (m_2^2 + m_1 m_2)}}{m_2^2 + m_1 m_2}.$$

This is far from elegant, but it does show mathematically that ϵ has an upper limit. This is because there is a $m_1^2 m_2^2 u^2 - 2\epsilon m_1 (m_2^2 + m_1 m_2)$ term under a square root sign. For the equation to work, this value has a lower limit of zero. The only variable in this expression is ϵ, so by setting $m_1^2 m_2^2 u^2 - 2\epsilon m_1 (m_2^2 + m_1 m_2) = 0$, the value of the *upper limit* for the kinetic energy loss is $\epsilon_{max} = \frac{1}{2} \left(\frac{m_1 m_2}{m_1 + m_2} \right) u^2$.

That there is an upper limit might seem surprising — at first thought you may be forgiven for believing all the kinetic energy could be lost in such a collision — but consider a moving object colliding with a stationary one: for all the kinetic energy to be lost, *both* objects would have to be stationary after the collision. This just doesn't happen — the objects always both keep moving at a slower speed.

The corresponding velocities for $\epsilon = \epsilon_{max}$ are the same and given by $v = w = \frac{m_1}{m_1 + m_2} u$.

As the velocities are the same, the masses effectively stick together. In fact, the best way to engineer a completely inelastic collision, which is quite common, is to make sure the objects stick together on collision.

10.1.2.1. *Instances of inelastic collisions*

Any macroscopic collision involving contact will involve some sort of energy dissipation as heat, sound or permanent deformation of one or both of the objects. If there is any sort of adhesion then the collision can be perfectly inelastic and a maximal amount of kinetic energy can be lost. Adhesion could be something colliding with a sticky object, or Velcro objects colliding, or a rugby tackle — when one person grabs onto another during their collision it is effectively a form of adhesion ensuring both players move at the same speed after the collision which is thereby completely inelastic.

10.1.2.2. *Reduced mass*

Notice that in a completely inelastic collision the kinetic energy lost is given by $\frac{1}{2}\mu u^2$ where μ has units of kilograms and is defined by $\mu = \left(\frac{1}{m_1} + \frac{1}{m_2}\right)^{-1} \equiv \frac{m_1 m_2}{m_1 + m_2}$ — i.e. it is the "product over sum" of the masses in the kinetic energy lost. This is known as the *reduced mass* of a two-body system. Although the term is not met again in this textbook, it is used quite often in certain aspects of the discipline.

Notice that μ is always less than both m_1 and m_2, is exactly half of one of the masses if they are equal and tends to the value of the *lower* mass if one of the masses is much heavier than the other.

10.1.3. *Superelastic Collisions*

If two objects collide and one or both of the objects releases some stored potential energy as a result of the collision which goes into *increasing* the kinetic energy of the system then the collision is said to be *superelastic*. This can be said to be an "explosion". In superelastic collisions:

(1) Total energy is conserved.
(2) Kinetic energy is increased.
(3) Linear momentum is conserved.

For a two-body collision, the linear momentum expression remains the same as the general case in Section 10.1 but the kinetic energy expression is modified to:

$$\frac{1}{2}m_1 u_1^2 + \frac{1}{2}m_2 u_2^2 + \epsilon = \frac{1}{2}m_1 v_1^2 + \frac{1}{2}m_2 v_2^2,$$

where ϵ is the (positive) energy added as a result of the collision.

The energy could be added as the result of a chemical or nuclear reaction (e.g. in a bomb), electrostatic or magnetic repulsion as the result of like charges or poles that are close together being liberated or a spring's repulsion.

10.2.   Reference Frames

Provided the observer's reference frame is inertial (i.e. not accelerating), the nature of a collision (elastic, inelastic or superelastic) is reference frame independent. Linear momentum will always be conserved and the kinetic energy change will always be the same for any inertial observer. In a two-particle collision it is natural to analyse the problem from the point of view of a reference frame moving at the same velocity of one of the particles before or after the collision to make the problem easily tractable. However, it is always possible to choose a reference frame in which the **total momentum of the system is zero**. This approach can often lead to a more intuitive grasp of the physics and make the mathematical manipulation simpler.

Consider a particle of mass m_1 travelling in a straight line at speed v_1 on a collision course with a particle of mass m_2 travelling in a straight line at speed v_2. The paths are at an angle α to each other. Let us investigate how the system can be studied by (a) considering the total momentum of the system to be zero and (b) considering the initial momentum of one of the particles to be zero and discuss the merits of each approach.

For part (a), consider the system such that the x-axis runs parallel to m_1 and the y-axis is perpendicular to this. Recognise that in a two-particle collision *we can always arrange the co-ordinate system axes to be anything we want*. It is sensible to keep things simple and choose our x-axis to be parallel to the path of one of the particles. The general picture is shown in Figure 10.5.

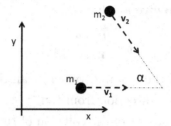

Figure 10.5: **Two particles on a collision course with an x-axis chosen along the original direction of one of the particles.**

To find a zero momentum frame of reference we need to find a single velocity to travel at that will give the sum of the two particles' momenta to be zero.

What **horizontal** component of velocity would we need to have so that the total horizontal momenta of the particles is zero?

Let the required horizontal component of velocity be v_x.

If we move along the x-axis at this speed, then the horizontal component of momentum of m_1 is $m_1(v_x - v_1)$ and the corresponding value for m_2 is $m_2(v_x - v_2 \cos \alpha)$.

Therefore, the condition for the total horizontal momentum to be zero is $m_1(v_x - v_1) + m_2(v_x - v_2 \cos \alpha) = 0$.

This rearranges to give a value of:

$$v_x = \frac{m_1 v_1 + m_2 v_2 \cos \alpha}{m_1 + m_2}.$$

What **vertical** component of velocity would we need to have so that the total vertical momenta of the particles is zero?

Let the required vertical component of velocity be v_y.

If we move along the y-axis at this speed then the vertical component of momentum of m_1 is $m_1 v_y$ and the corresponding value for m_2 is $m_2(v_y - v_2 \sin \alpha)$.

Therefore, the condition for the total vertical momentum to be zero is $m_1 v_y + m_2(v_y - v_2 \sin \alpha) = 0$.

This rearranges to give a value of:

$$v_y = \frac{m_2 v_2 \sin \alpha}{m_1 + m_2}.$$

The magnitude of the required velocity can therefore be obtained from $\sqrt{v_x^2 + v_y^2}$ and its direction from $\tan^{-1} \frac{v_y}{v_x}$.

Regardless of the nature of the collision of the particles (inelastic, elastic or superelastic), if we move at this velocity the total momentum of the system will always be zero before, during and after the collision process.

For part (b), consider the same system as given in Figure 10.5 but now consider that we ride alongside m_1 at velocity v_1.

In this case, relative to us, m_1 is stationary so it has zero momentum.

The horizontal component of m_2's velocity is now $v_2 \cos \alpha - v_1$, so the horizontal momentum of the system is $m_2(v_2 \cos \alpha - v_1)$.

The vertical component of m_2's velocity is now $v_2 \sin \alpha$, so the vertical momentum of the system is $m_2 v_2 \sin \alpha$.

10.3. Particle–Wall Collisions

Consider a particle striking a wall and rebounding at the same speed it moves in with, and with the angle of incidence equal to the angle of reflection, as in Figure 10.6.

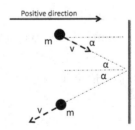

Figure 10.6: **A particle making an elastic collision with a wall with the angle of incidence equalling the angle of reflection.**

Vertically, the particle experiences no force by the wall and therefore no momentum change.

Horizontally, the change in momentum of the particle is:

$$\Delta p_{horizontal} = p_{after} - p_{before} = -mv\cos\alpha - mv\cos\alpha = -2mv\cos\alpha.$$

Hence, the average force *by* the wall *on* the particle is $\frac{-2mv\cos\alpha}{\Delta t}$ where Δt is the duration of the collision.

By Newton's third law, the force *by* the particle *on* the wall is $\frac{2mv\cos\alpha}{\Delta t}$ — i.e. from left to right on the figure.

If the particle did not rebound but became embedded in the wall then the factor of 2 would be dropped.

If the particle struck the wall at an angle of 90° to the surface, the factor of $\cos\alpha$ would be dropped (i.e. this force value is a maximum).

Now consider if a stream of particles makes a head-on collision with a wall and rebounds elastically (e.g. firing a stream of ping-pong balls against a door). Let the rate of particles passing a given point per unit time be R particles per second, and consider the situation in which the time between collisions, $\left(\frac{1}{R}\right)$, is much smaller than the duration of the collision. This will give an almost constant force against the wall.

The rate of change of momentum will be $2mv$ divided by the time between collisions, hence the force on the wall is $2mvR$.

If the ping-pong balls were to somehow embed in the door, the factor of 2 would again be dropped.

This principle is used to calculate the pressure of a gas in a container and is the basis of a standard calculation for the pressure of an ideal gas in the kinetic theory of gases.

10.4.   Fluid Jet Pressure

A jet of water fired against a wall effectively makes an ongoing inelastic collision with the wall, as demonstrated in Figure 10.7.

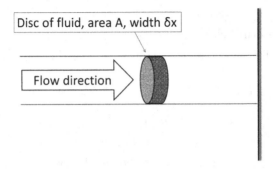

Figure 10.7:   A jet of fluid making an inelastic collision against a wall.

If the jet has velocity v, cross sectional area A and is incompressible with density ρ then the pressure with the wall is calculated as follows:

The volume of a thin element of fluid of width δx is $A(\delta x)$.

The mass of this element is $\rho A(\delta x)$.

The linear momentum of the element is $\rho A v(\delta x)$.

Hence, assuming the fluid *does not rebound*, the rate of change of momentum of the fluid is $-\frac{\rho A v(\delta x)}{(\delta t)}$ where (δt) is the increment of time over which the element impacts with the wall.

In the limit that the increment size and time tend to zero (i.e $\frac{\delta x}{\delta t} \to \frac{dx}{dt} = v$), the expression for the force *by* the wall *on* the fluid is given by $-\rho A v^2$.

By Newton's third law, the force by the fluid on the wall is $+\rho A v^2$ and hence the fluid pressure is given by:

$$P = \rho v^2, \tag{10.7}$$

i.e. the pressure is proportional to the velocity squared.

This is also the resistive force supplied by a stationary fluid with an object moving through it with velocity v and is the *Newtonian drag* mentioned in Subsection 5.2.6.

10.5. Rocket Propulsion

10.5.1. *The Basic Principle of Rocketry*

This section studies the basic dynamics of rocket propulsion. All rockets, regardless of their complexity, work by the conservation of linear momentum as applied to a system made up of (1) Rocket + (2) Fuel that gets ejected.

The rocket effectively works by superelastic collision — the vehicle travels at a certain velocity (and therefore kinetic energy) and increases this by using some form of stored (potential) energy to eject fuel out of the back which causes the rest of the rocket to move forwards faster by Newton's third law.

In a space rocket the potential energy is usually chemical–potential and is released by burning of the gases, while in a toy water bottle rocket the stored energy is mainly elastic potential in the walls of the bottle. A rather basic "rocket" is simply a balloon that is blown up and then let go; in this case it is also the stored elastic potential energy that works on the gas within the balloon.

To solve the system, Newton's second law in the form $F = ma$ may not be used as the mass under investigation is variable (as alluded to in Section 4.4).

10.5.2. *Rocket Propulsion for a Constant Velocity Fuel Ejection*

Consider a rocket in space (away from any gravitational fields) travelling at initial velocity u relative to a inertial reference frame. The rocket can eject fuel at a fixed speed v_0 *relative to the rocket*. Between a start time t and a final time $t + \delta t$, the rocket burns a mass δm of fuel.

The term δm requires some thought. The value itself will be positive and the change in mass of the ejected fuel will also be positive and equal to $+\delta m$. However, the change in mass of the *rocket* will be negative and equal to $-\delta m$.

The increase in velocity as a result of ejecting the fuel is δv.

The momentum of the system (rocket plus fuel) before ejection (time t) is $(m + \delta m)u$.

The momentum of the rocket after ejection (time $t + \delta t$) is $m(u + \delta v)$ and the momentum of the fuel is $\delta m(u - v_0)$, so the total momentum of the system is $m(u + \delta v) + \delta m(u - v_0)$.

Using the conservation of linear momentum this gives:

$$mu + (\delta m)u = mu + m(\delta v) + (\delta m)u - (\delta m)v_0,$$

where the brackets have been unwrapped.

Standard cancelling of terms gives $m(\delta v) = \delta m v_0$ so $\delta v = v_0 \frac{\delta m}{m}$.

The change in mass applies to the change in mass of the *ejected fuel*.

It can thus be written $\delta v = -v_0 \frac{\delta m}{m}$ where δm is now the change in mass of the rocket itself and will be negative, hence the change in velocity will be positive.

To find the velocity of the rocket as a function of its mass, the time increment can be made infinitesimally small and the resulting integral expression is $\int_u^v dv' = -v_0 \int_{m_i}^{m_f} \frac{dm}{m}$ where the limits refer to the final and initial values of the quantities.

This has the solution:

$$v = u - v_0 \ln \left(\frac{m_f}{m_i} \right) = u + v_0 \ln \left(\frac{m_i}{m_f} \right), \qquad (10.8a)$$

which is a standard equation for rocket propulsion with constant gas expulsion velocity.

If the final mass is instead written $m_f = m_i - m_e$, where m_e is the total mass of fuel ejected during the flight, then this can be rewritten:

$$v = u - v_0 \ln \left(1 - \frac{m_e}{m_i} \right), \qquad (10.8b)$$

which is a more tangible format.

For example, imagine a small rocket of mass without fuel of 1000 kg carrying a mass of fuel of 500 kg, so $m_i = 1500$ kg. If the rocket is at rest in space, then $u = 0$ and the specific versions of equations 10.8a and b are as follows:

For 10.8a, $v = v_0 \ln \frac{1,500}{m_f}$ where m_f starts at 1500 kg at the initiation of fuel ejection and finishes at 1000 kg when all the fuel is spent, giving a final velocity of $v = v_0 \ln \frac{1,500}{1,000} = v_0 \ln \frac{3}{2} \approx 0.4v_0$.

For 10.8b, $v = -v_0 \ln \left(1 - \frac{m_e}{1,500}\right)$ where m_e starts at zero and finishes at 500 kg when all the fuel is spent, giving a final velocity of $v = -v_0 \ln \left(1 - \frac{500}{1,500}\right) = -v_0 \ln \frac{2}{3} = +v_0 \ln \frac{3}{2}$, which is as before, as it should be.

Notice that for this idealised case of constant fuel ejection speed, the final rocket speed is directly proportional to the speed of fuel ejection, so this is of huge importance in rocket design. Space rockets currently have an exhaust speed of several kms^{-1}.

Of course, this is an idealised situation. If a rocket is escaping a gravitational field, an extra force term would need to be included, and if the fuel ejection speed varies, the differential equations become more complicated. Equation 10.8 is really for one of the simplest situations.

11

Motion on a Curved Path

This chapter studies the motion of a particle on a curved path. It starts by looking at uniform circular motion, then generalises the analysis to motion in a circle with changing speed and motion on any general curved path. The treatment here will be two dimensional for simplicity; generalisations to 3D are always possible in spirit (but may be tricky in practice).

11.1. Uniform Circular Motion

11.1.1. *General Kinematic Analysis*

Consider a particle moving along the arc of a circle of radius r at constant speed v. As previously stated (see Section 2.2), the particle must be accelerating as its velocity is constantly changing direction.

You probably know that in this case of constant speed, the acceleration of the particle is towards the centre of the circle; this can be logically reasoned in two sentences as follows:

(1) As the particle's speed stays constant the component of the acceleration of the particle in the direction of its motion *must* be zero.
(2) Hence, the acceleration of the particle must be along the vector joining it to the centre of the circle — i.e. if the particle is accelerating it can only be at right angles to the direction of motion along which it is *not* accelerating.

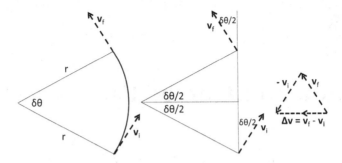

Figure 11.1: A particle moving along the arc of a circle of radius r at uniform speed from one point to another in a time interval δt with the velocity vectors at the start and end of the path and the change in velocity.

A geometric/mathematical analysis follows. Consider the particle's motion over an increment of time δt during which it subtends an angle $\delta\theta$ of the circle, as in Figure 11.1.

Let the speed of the particle be v at all times, with an initial velocity given by v_i and a final velocity v_f as indicated on Figure 11.1 with $v = v_i = v_i$.

The distance travelled in the short time is $v \cdot \delta t$ and the angle (in radians) between the corresponding directions of r is given by $\delta\theta \approx \frac{v.\delta t}{r}$.

Considering the change in velocity between the start and finish point:

The change in velocity *parallel* to the line joining the start and finish points (refer to the middle section of Figure 11.1) is $v\cos\frac{\delta\theta}{2} - v\cos\frac{\delta\theta}{2} = 0$ — i.e. there is *no change in velocity parallel to the line.*

The change in velocity *perpendicular* to the line joining the start and finish points is $v\sin\frac{\delta\theta}{2}$ *towards* the centre minus $v\sin\frac{\delta\theta}{2}$ *away from* the centre $= 2v\sin\frac{\delta\theta}{2}$ *towards* the centre of the circle.

The vector diagram on the right of Figure 11.1 indicates this result.

As the time interval tends to zero, then so does $\delta\theta$ and this change in velocity becomes $v \cdot \delta\theta$ because of the small angle approximation, i.e. $\sin \frac{\delta\theta}{2} \to \frac{\delta\theta}{2}$, so we can write $\delta v = v \cdot \delta\theta$. But as $\delta\theta \approx \frac{v.\delta t}{r}$, the expression for the change in velocity over an infinitesimal time is $\delta v = \frac{v^2.\delta t}{r}$ and hence the magnitude of the rate of change of velocity, $\lim_{t\to 0} \frac{\delta v}{\delta t}$ (the acceleration), is given by:

$$a = \frac{v^2}{r},$$ (11.1a)

with the direction always being towards the centre of the circle, i.e. radially inward. This is sometimes called the *centripetal acceleration* of the particle. There is a need for a component of centripetal acceleration for any motion which is not along a straight line, as any change in path implies a change in velocity perpendicular to the velocity itself.

Though angular velocity, ω, will not be defined formally until Chapter 14, it is worth also mentioning here that two other equivalent formulae can be used for acceleration in a circle at uniform speed, namely:

$$a = \omega^2 r,$$ (11.1b)

and the less-often seen and used

$$a = v\omega.$$ (11.1c)

11.1.2. *What This Tells Us*

The kinematic analysis plus knowledge of Newton's second law means that 1) if we analyse the dynamics of an object and decide that the resultant external force on it is perpendicular to its velocity then it must (for that instant at least) be following a circular path; and, conversely, 2) if we see an object moving in a circular path we know that the resultant external force on the object must be perpendicular to its velocity.

This is another example of deductive vs. inductive reasoning (as mentioned previously in Section 4.5).

Nature is replete with examples of objects travelling with uniform circular motion. We shall study several as part of this book with one in detail here.

11.1.3. *Example of An Object Travelling Around a Circular Banked Track*

Consider an object of mass m travelling around a circular banked track at constant speed, such as a cyclist in a velodrome. The forces acting on the cyclist will be the weight, the contact force and friction. The direction of the friction will depend on how fast the cyclist is going — the free body diagram in Figure 11.2 shows the condition for when the cyclist is going quick enough such that he or she has a tendency to slide *up* the slope and the friction opposes this motion. If the motion were slow, he or she would tend to slide *down* and the friction would act up the slope.

The motion is analysed by splitting the forces into components:

Vertically, the system *is in equilibrium* so the upward component of the contact force must equal the weight plus the downward component of the friction, i.e.

$$mg + F\sin\theta = N\cos\theta. \qquad (11.2)$$

Horizontally, the object *is accelerating towards the centre of the track* with magnitude $\frac{v^2}{R}$. This acceleration will be supplied by the

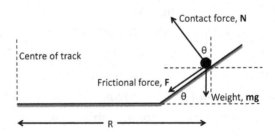

Figure 11.2: Free body diagram for a fast object on a banked track — the frictional force acts down the slope to prevent the object moving upward.

horizontal components of the contact force and the friction, so using $F = ma$ gives:

$$\frac{mv^2}{R} = N \sin \theta + F \cos \theta. \tag{11.3}$$

We now have two equations with two unknowns, N and F. By substitution and a little algebra, this leads to expressions for the friction and contact forces of:

$$\text{Frictional force } F = m \left(\frac{v^2}{R} \cos \theta - g \sin \theta \right),$$

$$\text{Contact force } N = m \left(\frac{v^2}{R} \sin \theta + g \cos \theta \right).$$

These results merit some discussion:

- For a general angle, the frictional force will be positive if $\frac{v^2}{R} \cos \theta > g \sin \theta$ and negative (i.e. contrary to the direction on the free body diagram) if $\frac{v^2}{R} \cos \theta < g \sin \theta$.

  This corresponds to the cyclist going fast (as represented by the free diagram), or going slow, when the friction must act the other way. The solution was designed for the former case but also works for the latter.

- If $\frac{v^2}{R} \cos \theta = g \sin \theta$ then the frictional force will be exactly zero. This defines the speed $v = \sqrt{gR \tan \theta}$ at which an object would be able to slide around the track if it were frictionless.

- No matter what the angle, the frictional force can always reach a maximum value that depends on the two materials (see Section 5.2.3 — $F_{max} = \mu N$). If it exceeds this value then the object will slide up or down the track and the equations will be rendered invalid.

- Setting an angle of zero corresponds to flat ground; the contact force equals the weight ($N = mg$) and the friction is the sole supplier of the centripetal acceleration ($F = m\frac{v^2}{R}$).

- Setting an angle of 90° corresponds to a "wall of death"; the friction equals the weight ($F = -mg$) and the contact force is the sole supplier of the centripetal acceleration ($N = m\frac{v^2}{R}$).

11.2. Motion on a General Curve with Changing Speed

If a particle changes *speed* as it travels along a circular path it has, in addition to its centripetal acceleration towards the centre of curvature, a component of acceleration at a tangent to the path. The tangential component represents the rate of change of *magnitude* of the velocity vector, i.e. the rate of change of *speed*. These two accelerations are perpendicular and are therefore given by:

$$\text{Radial component (perpendicular to velocity): } a_{radial} = \frac{v^2}{r}$$

$$\text{Transverse component (parallel to velocity) } a_{tranverse} = \frac{d|\boldsymbol{v}|}{dt}$$

So if an object travels in a circle with *increasing* speed it does not accelerate towards the centre, rather just "forward" of the centre and if decreasing in speed just "backward" of the centre.

This applies to any curved path. For any point along the path there will be a centre of curvature and associated radius of curvature, which will in general change along the path. Provided r and v are the *instantaneous* values of radius of curvature and speed then the above expressions are perfectly general.

11.2.1. *More on the General Radius of Curvature and How to Use it with the Circular Motion Equation*

If a curve is given in Cartesian coordinates by $y(x)$ then the *radius of curvature* of the curve at *any* point is given by:

$$R = \frac{\left(1 + \left(\frac{dy}{dx}\right)^2\right)^{\frac{3}{2}}}{\frac{d^2y}{dx^2}}.$$

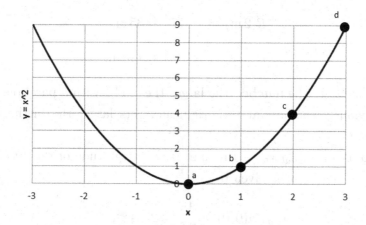

Figure 11.3: An object at four different locations on a parabolic path. The radius of curvature increases with increasing x. The axes are not drawn to scale.

(No proof is provided for this — that would be the job of a geometry textbook — and not many physicists will have it committed to memory, but it is useful to know the formula exists and where to look it up.)

This means that if a particle has a known speed and acceleration and travels on a path that can be described by a differentiable function then the acceleration can always be found exactly.

For example, consider a perfectly parabolic path defined by $y = x^2$ as shown in Figure 11.3.

As $y = x^2$ then $\frac{dy}{dx} = 2x$ and $\frac{d^2y}{dx^2} = 2$, so the radius of curvature at any point is given by $\frac{(1+(2x)^2)^{\frac{3}{2}}}{2} = \frac{1}{2}(1 + 4x^2)^{\frac{3}{2}}$.

If the particle travels at a constant speed of $1\,\text{ms}^{-1}$ then what is the magnitude of its acceleration at, say, $x = 0, 1, 2$ and $3\,\text{m}$?

The radius of curvature at the four points is:

$$R(0) = 0.5\,\text{m},$$

$$R(1\,\text{m}) = \frac{1}{2} \cdot 5^{\frac{3}{2}} \approx 5.6\,\text{m},$$

$$R(2\,\mathrm{m}) = \frac{1}{2} \cdot 17^{\frac{3}{2}} \approx 35\,\mathrm{m},$$

$$R(3\,\mathrm{m}) = \frac{1}{2} \cdot 37^{\frac{3}{2}} \approx 113\,\mathrm{m}.$$

The larger the number, the larger the circle; with a parabola this is obviously smallest at the origin and gets larger the further one moves away.

So using $a_{radial} = \frac{v^2}{r}$, this means the magnitude of the acceleration at each point is given by:

$$a_{radial}(0,0) = \frac{1^2}{0.5} = 2\,\mathrm{ms}^{-2},$$

$$a_{radial}(1,1) = \frac{1^2}{5.6} = 0.18\,\mathrm{ms}^{-2},$$

$$a_{radial}(2,4) = \frac{1^2}{35} \approx 0.029\,\mathrm{ms}^{-2},$$

$$a_{radial}(3,9) = \frac{1^2}{113} \approx 0.009\,\mathrm{ms}^{-2}.$$

This corroborates something anyone who has ever driven a car or a bicycle, or even run along the ground knows well: the tighter the arc of a bend, the greater the acceleration, meaning more effort needs to be put in to maintain the curved path.

The direction of the acceleration can also be found easily at any point — it is simply perpendicular to the tangent to the curve at that point.

If the object is also experiencing a rate of change of speed then the same analysis applies for the radial acceleration, and to get the total acceleration the two perpendicular components can be added as vectors.

11.2.2. *Example of an Object Sliding Off a Round, Frictionless Hill*

Consider an object on top of a hemisphere as shown in Figure 11.4.

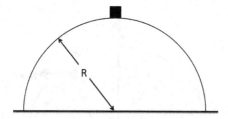

Figure 11.4: **An object in unstable equilibrium on top of a frictionless hemisphere.**

If the object is placed exactly at the apex of the hemisphere and is perfectly still then it will remain there ad infinitum. If, however, it is ever so slightly to one side of the apex, or is moving at the tiniest velocity then the object will slide around the slope. Such an object is said to be in *unstable equilibrium*.

It seems likely that when the object slides around the hemisphere it will not remain in contact with the hemisphere all the way round, thus will lose contact with it at some point. The main problem here is to determine at what point.

The object will slide around the hemisphere. As no non-conservative forces dissipate any energy, the kinetic energy gained by the object equals the gravitational potential energy it loses. It therefore gets faster. The acceleration can be split into two components at any time:

(1) The component towards the centre of the shape, which is supplied by the component of the object's weight towards the centre of the shape minus the contact force.

(2) The component at a tangent to the shape which is supplied entirely by the component of the weight parallel to the surface at that point.

These can be deduced with the aid of a free body diagram for the object, shown in Figure 11.5 at three points: the top, a general distance around and just before leaving the surface.

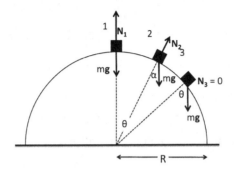

Figure 11.5: **Free body diagram for the object in Figure 11.4 at the start, middle and end of its contact with the shape.**

At point 1, $N_1 = mg$, so there is no acceleration but a slight disturbance from equilibrium causes movement.

At point 2, the object is accelerating such that $ma_{tranvserse} = m\frac{d|\mathbf{v}|}{dt} = mg \sin \alpha$ and $ma_{radial} = m\frac{v^2}{R} = mg \cos \alpha - N_2$.

The latter equation demonstrates that the contact force *must* fall as the speed increases. Rearranging, we can write $N_2 = m(g \cos \alpha - \frac{v^2}{R})$. As α increases, $\cos \alpha$ decreases, v^2 increases and m, g and R are constant. This is probably what one would intuitively expect.

At point 3, the contact force has fallen to zero and at the moment the object leaves the shape the acceleration towards the centre is supplied entirely by the component of the weight in that direction. This gives an expression for the component of acceleration at this point of $\frac{v^2}{R} = g \cos \theta$.

The velocity can be found by the conservation of energy, plus a little geometry, and by equating the kinetic energy gained to gravitational potential energy lost. When the sliding object is at an angle θ, the height below the vertical the object has dropped is given by $R(1 - \cos \theta)$, and thus the potential energy lost is $mgR(1 - \cos \theta)$. Therefore, $\frac{1}{2}mv^2 = mgR(1 - \cos \theta)$, hence $\frac{v^2}{R} = 2g(1 - \cos \theta)$.

Notice that the contact force does no work on the object (as it is always perpendicular to its direction of motion) and thus the total mechanical energy (kinetic plus potential) does not change.

So we now have two expressions for $\frac{v^2}{R}$ which can be equated to give $g\cos\theta = 2g(1 - \cos\theta)$.

Cancelling g and rearranging this gives a value of the angle of departure of $\cos^{-1}\frac{2}{3} \approx 48°$. This is a neat result, which rather counter-intuitively neither depends on g nor R. It does not matter how big the sphere is, and what strength of uniform gravitational field the scenario is in, the object always leaves the sphere at the same angle.

If friction is present then the object still leaves when $\frac{v^2}{R} = g\cos\theta$, according to the analysis at point 3. There is no reason for this part of the analysis to change. The speed at this point will be different, however, as some of the potential energy will have been dissipated by friction during sliding. In the absence of any other information about the nature of the frictional force, all that can be said is that it will have caused a positive amount of energy to be dissipated, which we can designate by ϵ.

The energy expression then becomes $mgR(1 - \cos\theta) = \frac{1}{2}mv^2 + \epsilon$, hence $\frac{v^2}{R} = 2g(1 - \cos\theta) - \frac{2\epsilon}{mR}$. This now gives $\frac{v^2}{R}$ terms which equate to $g\cos\theta = 2g(1 - \cos\theta) - \frac{2\epsilon}{mR}$, which rearrange to give the modified expression $\theta = \cos^{-1}\left(\frac{2}{3} - \frac{2\epsilon}{3Rmg}\right)$.

As ϵ is positive, this means $\frac{2}{3} < \left(\frac{2}{3} - \frac{2\epsilon}{3Rmg}\right)$, and thus the angle of departure is greater when energy is dissipated, as would probably be expected. This analysis also fixes the maximum amount of energy that can be dissipated during the slide, which occurs when $\theta = 90°$ and $\cos\theta = 0$ when $\left(\frac{2}{3} - \frac{2\epsilon}{3Rmg}\right) = 0$, so $\epsilon_{max} = Rmg$.

Another interesting variant of the problem involves an object rolling from the apex rather than sliding. Though the problem is not analysed in detail, you will be equipped to investigate this yourself after Chapter 17.

12
Simple Harmonic Motion

This chapter looks at oscillatory motion, in particular oscillations where the displacement of an object is directly proportional to the restoring force. This is known as *simple harmonic motion* (SHM) and is a common natural phenomenon. It is investigated using both a force and a potential energy approach.

This chapter covers only the basics of what is really a large topic in its own right. University-level courses on vibrations and waves pick up where pure mechanics leaves off; two notable topics that are omitted here are resonance and damping. The level of mathematical sophistication is also increased as an extensive use of complex analysis is usually invoked to treat waves' behaviour.

12.1. Amplitude, Period, Frequency and Angular Frequency

A (mechanical) oscillation is any repetitive motion about a central point. Before studying the physics of oscillations, it is prudent to define some general quantities relating to them. These definitions are used consistently within this text and closely match those used most frequently in contemporary courses in physics in the UK. Note that some of the definitions can be slightly different in discourses outside of this range, so take care.

The **amplitude** of an oscillation, A, is defined as the maximum magnitude of the displacement of the oscillator from equilibrium. It is a positive scalar with SI units of metres.

The **period** of an oscillation, T, is the time taken for an oscillation to go through one complete cycle. It is a positive scalar with SI units of seconds.

The **frequency** of an oscillation, f, is defined as the number of oscillations per unit time. It is a scalar with derived SI units of hertz (Hz) (or base units of s^{-1}). It is related to period by:

$$f = \frac{1}{T}. \tag{12.1}$$

The **angular frequency** of an oscillation, ω, is given by:

$$\omega = 2\pi f = \frac{2\pi}{T}. \tag{12.2}$$

It is a scalar with units of $rad.s^{-1}$ where the rad refers to the angular unit the radian. There is more discussion on the radian in Chapter 14.

12.2. Sinusoidal Oscillations

Consider a particle oscillating sinusoidally with amplitude A and angular frequency ω. Its displacement, x, as a function of time, t, can be written:

$$x = A\sin(\omega t). \tag{12.3a}$$

The velocity, v, of the particle as a function of time will therefore be:

$$v = \omega A\cos(\omega t), \tag{12.3b}$$

and its acceleration, a, will be given by:

$$a = -\omega^2 A\sin(\omega t). \tag{12.3c}$$

It can be seen therefore that the acceleration of a sinusoidal oscillator can be expressed succinctly by:

$$a = \frac{d^2x}{dt^2} = -\omega^2 x. \tag{12.4}$$

Any oscillator that follows this pattern is known as a **simple harmonic oscillator**. They have the following characteristics:

(1) The amplitude is constant (this is what is meant by "simple").
(2) The frequency and period are independent of amplitude.
(3) The time dependence of the displacement is a sinusoid of a single frequency.

If a mechanical system exists such that the acceleration of an object is proportional to its displacement then it will be a simple harmonic oscillator with a period defined by Equation 12.4.

If this is true *and* there is a position of stable equilibrium for the system *and* there are no dissipative forces causing energy losses in the system then the object will undergo simple harmonic motion (SHM).

12.2.1. *A Simple Harmonic Oscillator Does not Necessarily Exhibit SHM*

If a person took a pen and with some considerable skill moved it backwards and forwards about a point according to Equation 12.4 then the pen would exhibit a simple harmonic oscillation but would not be undergoing simple harmonic motion. Confusing? The issue is that the pen would not obey the criteria outlined in the previous paragraph: there would be no point of stable equilibrium for the pen and the pen's system would be entirely governed by non-conservative and thus dissipative contact forces.

This example would be difficult to realise with much accuracy but machines with cam systems produce such behaviour when the cam rotates at a steady rate. While the mathematics and the kinematic aspects of the physics that apply to simple harmonic oscillators work for these systems, the details of the energy of the system (see Section 12.5) simply cannot be used.

12.3. Two Examples of SHM

There now follow two examples of mechanical systems that exhibit SHM. They highlight the usual procedure for the identification of SHM by analysis of the dynamics of a system.

1. Spring displaced from equilibrium

Consider a spring in a frictionless environment that supplies a restoring force to a mass m that is proportional to its displacement from the equilibrium position and acts in the opposite direction to the displacement, as shown in Figure 12.1.

Unless the spring is stationary at the equilibrium point then it will oscillate about this point. Wherever it is, regardless of its velocity, it will accelerate back to this equilibrium point, and the further it moves away, the greater this acceleration will be. If the spring constant is given by k then applying Newton's second law to the mass gives:

$$m\frac{d^2x}{dt^2} = -kx,$$

and hence

$$\frac{d^2x}{dt^2} = -\left(\frac{k}{m}\right)x.$$

Comparing this with Equation 12.4 tells us that the mass oscillates with SHM with an angular frequency given by $\omega^2 = \frac{k}{m}$ — i.e.

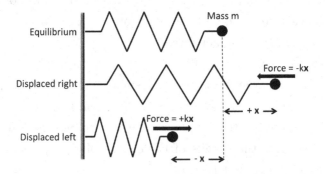

Figure 12.1: A spring at three different positions.

the period of oscillation is:

$$T = 2\pi \sqrt{\frac{m}{k}}. \tag{12.5}$$

Notice that this is *independent of amplitude.*

If the initial amplitude is A then the displacement, velocity and acceleration as a function of time are given by:

$$x = A \sin \left(\sqrt{\frac{k}{m}} t \right),$$

$$v = A \sqrt{\frac{k}{m}} \cos \left(\sqrt{\frac{k}{m}} t \right),$$

$$a = -\frac{k}{m} A \sin \left(\sqrt{\frac{k}{m}} t \right).$$

Notice that when the displacement is a maximum, velocity is zero and acceleration is a minimum. When the displacement is zero, the acceleration is also zero and the velocity is either maximum or minimum. Finally, when displacement is a minimum, the acceleration is a maximum and the velocity is zero.

These are general features of *all* simple harmonic oscillators.

2. Simple pendulum

Consider a simple pendulum of length l with a small mass m hanging in a uniform gravitational field. It is displaced a small distance x along the arc from its equilibrium position, as shown in Figure 12.2.

When in static equilibrium (i.e. simply hanging without movement or displacement), the tension equals the weight of the pendulum. But when moved to a certain amplitude from the equilibrium point and then released (from rest) the pendulum will move in a circular arc.

Initially there is no centripetal acceleration so the tension equals the component of the weight in the line of the string, $mg \cos \theta$. The

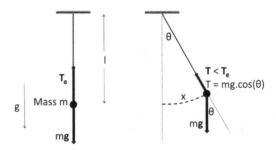

Figure 12.2: **The simple pendulum with a small displacement from its equilibrium position.**

transverse force will therefore be equal to $mg \sin \theta$ roughly towards the equilibrium point (actually at a tangent to the arc).

The equation of motion along the arc can be written $m\frac{d^2x}{dt^2} = -mg \sin \theta$, so $\frac{d^2x}{dt^2} = -g \sin \theta$.

The angle θ can be expressed by $\theta = \frac{x}{l}$ (this is the definition of angle in radians), so:

$$\frac{d^2x}{dt^2} = -g \sin \frac{x}{l}.$$

This *non-linear, second-order, ordinary differential equation* describes the motion of the pendulum bob exactly, but is not generally soluble. Try using an online solver to see.

However, if the angle $\theta = \frac{x}{l}$ is *small* (i.e. $l \gg x$) then $\sin \frac{x}{l} \approx \frac{x}{l}$ and the equation becomes:

$$\frac{d^2x}{dt^2} \approx -\frac{g}{l}x.$$

This is the equation of motion for small oscillations, and therefore the pendulum will exhibit simple harmonic motion with an angular frequency given by $\omega^2 = \frac{g}{l}$ — i.e. the period of oscillation is:

$$T = 2\pi \sqrt{\frac{l}{g}}, \tag{12.6}$$

which is a well-known result.

12.3.1. *What Does "Small Angle" Mean?*

As stated, the small angle approximation means $\theta \ll 1$ (i.e. $l \gg x$), but if we were carrying out an investigation using a simple pendulum, to what angle exactly is the approximation valid? Essentially it depends on the required accuracy of the final results. If, say, 10% accuracy is required then *very roughly* a ratio of $\frac{x}{l} = 0.1$ would be acceptable, and so on. A full check of this requires an in-depth uncertainty analysis and in practical circumstances would really require an empirical calibration of the pendulum. But the rule of thumb is fine for starting purposes and the principle that the smallness of the angle depends on the accuracy required is always valid.

In fact, the reader is urged to try an easy experiment now to see how valid the approximation is. Would you expect the period to get larger or smaller with an increasing angle?

12.4.   SHM and Uniform Circular Motion

If an object moves in a circle at constant speed then the x- and y-components of its motion both undergo SHM. This means if an object moves in such a manner in a plane, and the observer is looking along the plane (and has no depth perception) then the object will appear to move with SHM.

12.5.   Energy in SHM

12.5.1.   *Kinetic and Potential Energies*

When an object undergoes SHM without friction, its total energy remains constant. It always has a potential energy as a result of its position in the conservative field that exerts the conservative force on it, plus a kinetic energy due to its velocity. These are constantly changing but always add to give the same overall value, which will equal the energy supplied to the system in the first place to commence the oscillations.

The **kinetic energy** is given by:

$$E_K = \frac{1}{2}mv^2 = \frac{1}{2}m\omega^2 A^2 \cos^2(\omega t)$$

and has a maximum of $\frac{1}{2}m\omega^2 A^2$ when the particle passes through the equilibrium point.

The **potential energy** is given by $U = \frac{1}{2}kx^2$ where k is a constant (more on this constant later), so:

$$U = \frac{1}{2}kA^2 \sin^2(\omega t),$$

which has a maximum of $\frac{1}{2}kA^2$ when the particle reaches the extremes of the oscillation.

The **total energy** is then the sum of the two and is given by:

$$E_{total} = \frac{1}{2}m\omega^2 A^2 \cos^2(\omega t) + \frac{1}{2}kA^2 \sin^2(\omega t) = \frac{1}{2}kA^2$$

at all times (as $\omega^2 = km$ and $\sin^2 \omega t + \cos^2 \omega t = 1$).

The energies as a function of displacement and time are shown in Figure 12.3.

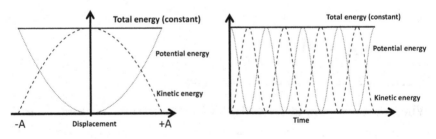

Figure 12.3:   Kinetic, potential and total energy as a function of displacement and time for a simple harmonic oscillator. The energy with time plot covers two complete oscillations starting from rest and maximum amplitude.

12.5.2. *The Constant, k*

For the spring example, the constant k is easily identified as the spring constant. For the pendulum, by using a little geometry and the small angle approximation it can be seen that the gravitational potential energy as a function of x is given by $\frac{1}{2}\left(\frac{mg}{l}\right)x^2$.

This means for a pendulum, the effective "spring constant" is equivalent to $\frac{mg}{l}$. For a spring, the spring constant is a measure of the spring's strength in pulling the spring back to equilibrium. Similarly, for a pendulum the quantity $\frac{mg}{l}$ is a measure of the pendulum–Earth system's strength in pulling the bob back to equilibrium.

All simple harmonic oscillators have this "effective spring constant".

12.5.3. *The Potential Well Approach*

The plot of potential energy vs. displacement in Figure 12.3 is a potential well of the form seen in Figure 9.5.

Though this is for stretching a spring, the same plot can apply to *any* simple harmonic oscillator as can be inferred from the note on the constant k above.

The negative gradient of the force gives the value of the spring constant, and as $\omega^2 = \frac{k}{m}$ this gives $\omega^2 = \dfrac{-\frac{dF}{dx}\big|_{at\,equilibrium}}{m}$.

Using $F(x) = -\frac{dU}{dx}$ provides an alternative equation for the 1D oscillator, viz.:

$$\omega^2 = \frac{\frac{d^2U}{dx^2}\big|_{at\,equilibrium}}{m}. \tag{12.7}$$

This equation provides an alternative method of deducing the period of SHM using an energy approach: if the second derivative is independent of x at the equilibrium point then the motion must be SHM and have an angular frequency given by Equation 12.7, and the period can be taken using Equation 12.2.

12.5.4. *Example with the Simple Pendulum Revisited*

For the pendulum in Figure 12.2, the potential energy as a function of displacement is exactly (i.e. with no approximations) $U = mgl(1 - \cos\theta) = mgl\left(1 - \cos\frac{x}{l}\right)$ (prove it!).

Differentiating once with respect to x gives $\frac{dU}{dx} = mg\sin\frac{x}{l}$, and differentiating again gives $\frac{d^2U}{dx^2} = \frac{mg}{l}\cos\frac{x}{l}$.

This is another example of a second-order non-linear differential equation without a solution. But near the equilibrium point of $x = 0$, the cosine term is approximately unity (this is the small angle approximation again), hence the second derivative of the potential energy is approximately constant about this point and can be written $\frac{d^2U}{dx^2} \approx \frac{mg}{l}$.

Therefore, using Equation 12.7 gives $\omega^2 = \frac{\frac{mg}{l}}{m} = \frac{g}{l}$ — i.e. the motion is simple harmonic with a period given by Equation 12.6 earlier.

12.6.   Other Features of SHM

There are many other important and interesting features of SHM (most notably *resonance* and *damping*) that we will not explore here, but you are now in a position to start learning. They typically form a major part of vibrations and waves courses at university level. For most of these courses, a decent knowledge of complex analysis is essential, so it is important to make sure that you have a good grasp of this part of mathematics before starting.

13

Gravitation

This chapter studies Newton's law of universal gravitation. It states the law, shows how it can be used for macroscopic bodies and highlights an important feature of the definition of mass. It looks at field strength and potential and shows a few important examples of how the law can be used. Though Einstein's theory of general relativity is not covered, some of the concepts that lead to its use are highlighted. The section finishes with some comments on the variation of the gravitational field strength on the Earth.

13.1. Newton's Law of Gravitation

In the late 17th century, Isaac Newton (having studied the work of Johannes Kepler and Tycho Brahe and "occasioned by the fall of an apple") realised that, in conjunction with his three laws of motion:

(1) Any two masses, m_1 and m_2 attract each other with a force of magnitude F that is proportional to the product of these masses i.e. $F \propto m_1 m_2$.

(2) This force is inversely proportional to the square of the distance, r, between the objects, i.e. $F \propto \frac{1}{r^2}$.

Combining these equalities and combining with a proportionality constant leads to Newton's law of gravitation:

$$F = -G\frac{m_1 m_2}{r^2}. \tag{13.1a}$$

G is the *gravitational force constant*, and is shown experimentally to be $G \approx 6.674210 \times 10^{-11}$ Nm^{2}kg^{-2}. The negative sign in the formula is the convention to show *attraction* between the particles. Some find it easiest to remember the approximate numerical value of the constant as $\frac{2}{3} \times 10^{-10}$.

The law can be written in vector format: if $\hat{r}$ is a *unit vector* joining the two masses then Equation 13.1a can be written:

$$\boldsymbol{F} = -G\frac{m_1 m_2}{r^2}\hat{r}. \tag{13.1b}$$

If **r** is the displacement between the two masses then:

$$\boldsymbol{F} = -G\frac{m_1 m_2}{r^3}\boldsymbol{r}. \tag{13.1c}$$

The law implies that *all* masses in the universe attract *all other masses* with a force given by Equation 12.1. This is well verified by experiment, notably by Cavendish in 1798.

13.1.1.   *The Gravitational Force is Weak*

As a consequence of G being such a small value, gravitational forces are only significant for planetary size masses. However, as masses are always positive, the forces are always attractive and add vectorially to give a large net force.

To develop an appreciation for the size of the force, consider two electrons separated by a certain distance. They will experience an electrostatic force of repulsion given by Coulomb's law and a gravitational force of attraction given by Equation 13.1. If you compute the relative magnitudes of the forces (try it!) you will find the electrostatic force of repulsion is significantly greater than the gravitational attraction. In fact, the ratio of the magnitudes of the forces is of the order of 10^{42}. This is a big number to encounter in the physical sciences in any context and is only beaten into second place in this book by the mass of the universe estimate in Chapter 1.

Luckily, this force is relatively weak; life would be somewhat difficult otherwise. There exists a gravitational force between you and every object you can see around you, but the force is near negligible. If it was not then stuff would tend to clump together. For most humans, the only gravitational force we directly experience is the uniform gravitational force between the Earth and ourselves.

13.1.2. *Point Masses*

Newton's law of gravitation applies to *point masses* only. The masses in Equation 13.1 effectively have all their mass concentrated at one point. In terms of its gravitational effect, *any* macroscopic object can be considered to be a point mass with its focus at the object's *centre of mass* (see Chapter 14).

For spherically symmetric bodies (e.g. planets), the centre of mass is at the centre of the object. That this behaves as a point mass is addressed in Section 13.4 and used in an example now.

13.1.3. *Example: Circular orbits about a planet (with a preface on Newton's cannon)*

Find the speed, v, necessary to orbit a distance r from a planet's centre in a *circular* orbit, plus the time taken for one complete orbit. The mass of the planet is M.

Newton's cannon

Consider the trajectory a ball takes when a person throws it with an initial velocity parallel to the ground. Let us make the situation hypothetical in that the planet the person throws the ball on has no atmosphere so there is no air resistance. Once the ball leaves the hand the only force on the ball is the gravitational force by the planet.

For a normal throw, the ball follows a parabolic trajectory (as many readers will have investigated in A-level mathematics) before striking the ground.

Now let us make the situation even more hypothetical by allowing the person to have superhuman strength such that they can throw the ball with any velocity they desire. The harder they throw the ball, the farther away it will land. If they throw it hard enough then the ball will "disappear over the horizon" before hitting the ground. This disappearing over the horizon corresponds to the ball no longer travelling over "flat" ground but in fact is following the curvature of the Earth. The force can no longer be considered as simply "downwards" for these prodigious throws, but rather towards the centre of the Earth. So if this superhuman threw the ball hard enough it would travel half way around the world before landing, and if they threw it harder still it would, in principle, travel all the way around the world before hitting the person on the back of the head! This corresponds to a perfectly circular orbit, where the radius of the orbit is equal to the radius of the planet plus the height above the surface that the ball was thrown from. If the superhuman happened to be of normal height then this would be almost the same as the radius of the Earth.

Let us consider this orbit in more detail: what happens is that as the ball travels, the force of gravity pulls it downwards by an amount which exactly matches the curvature of the Earth. There is one unique velocity that will do this for any given height: any slower and the ball will spiral in towards the planet; any faster and the ball will spiral away.

Newton outlined an argument similar to this when discussing orbits but using the example of a cannon and ball, hence this discussion is equivalent to that which is usually known as Newton's cannon.

Solving the problem at hand

Back to the problem, now recognising that one unique speed of orbit should follow from the preceding discussion.

The free body diagram for the orbiting object is shown in Figure 13.1.

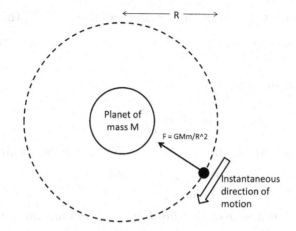

Figure 13.1: **An object moving in a circular orbit about a planet.**

Assuming the planet behaves as a point mass with the mass at its centre:

The object moves in a circle at constant speed so using Newton's second law of motion and $a = \frac{v^2}{r}$ for circular motion, if the object has a mass m then the resultant external force on the object is given by $\frac{mv^2}{r}$.

This force is solely due to the gravitational force of the planet, so using Newton's law of gravitation, $\frac{mv^2}{r} = \frac{GMm}{r^2}$.

Rearranging gives an expression for the orbital speed of:

$$v = \sqrt{\frac{GM}{r}}.$$

As expected, this shows that there is one specific velocity that is required for an orbit at a specific height: any faster and object will move away (gaining potential energy and losing kinetic energy); any slower and the object will spiral into the Earth (gaining kinetic and losing potential energy).

For a *near-Earth orbit* (i.e. just above the atmosphere) with $M_E = 6.0 \times 10^{24}$ kg and $R_E = 6.4 \times 10^6$ m, this gives an orbit velocity of $7.9\,k\text{ms}^{-1}$.

As the object travels in a circle and $speed = \frac{distance}{time}$, the equation for velocity can be written $\frac{2\pi r}{T} = \sqrt{\frac{GM}{r}}$ where T is the period of the orbit.

This rearranges to give:

$$T^2 = \frac{4\pi^2}{GM}r^3.$$

This important result shows that for a circular orbit about a planet, the *period squared is proportional to the radius of the orbit cubed*.

This is known as *Kepler's third law of planetary motion* and was one of the precursors of Newton's theory (though Kepler discovered the law empirically using data for planets about the Sun rather than satellites about a planet).

For a near-Earth orbit, this gives a period of 5090 s $\cong$ 0.059 days, which is equivalent to roughly 17 revolutions per day.

If an object is in orbit about a planet then the only force on it is the weight and the object is said to be in "free fall".

13.1.4.   *The Inaccuracy of the Term "Weightless"*

It is sometimes said that when an object is in orbit about a planet, then it is "weightless". Though this term is firmly within the English language, it is one of the worst imaginable to describe a phenomenon. The fact is, for an object to be in orbit, it *must* have weight. Indeed, for an object in a simple orbit described above, the *only force on the object is the weight*.

The reason that the term has arisen is because if an astronaut is inside a spacecraft in orbit about a planet then the normal contact forces that usually exist when the spacecraft is accelerating towards or away from a planet, or parked on a planet or moon, are zero. The analysis that applies to a single object in the example above applies to every object in the spacecraft, so the astronauts and all their belongings in the spacecraft appear to be floating. In fact, they are all in free fall. A better, though admittedly linguistically clumsy term for the phenomenon would be "contact forcelessness".

One strange fact that becomes apparent through all this though is that one never actually feels one's weight; rather we feel the contact force that acts in opposition to the weight. If the weight acts on its own then it is impossible to detect.

13.2. Gravitational Field Strength

13.2.1. *Gravitational Field Strength and Weight*

The weight of an object is the total gravitational force exerted on the object by all other masses in the universe.

When the object is near the surface of the Earth, all other masses can be neglected as all other stars, planets and moons are too far away to exert a measurable gravitational influence. The same applies if an object nears other planetary bodies. If this statement seems confusing, you are advised to look up some astronomical data on the distance to and masses of local celestial bodies and use Equation 13.1 to see for yourself how small the effects are.

Because an object's weight on Earth is given by mg this can be equated to Newton's law with the Earth's mass and radius, giving:

$$g = -G\frac{M_E}{R_E^2}. \tag{13.2}$$

This is known as the *gravitational field strength* of the Earth. Its value varies from place to place on the Earth but is approximately $9.8\,\text{Nkg}^{-1}$ — this is the force per unit mass of any object near the Earth's surface.

The same analysis could, of course, be applied to any other massive body.

13.2.2.   *g: Gravitational Field Strength in Nkg^{-1} or Acceleration Due to Gravity in ms^{-2}?*

When thinking of g in terms of gravitational fields and forces, it is usually better to think of it as the former, and when considering the kinematics near the surface of the Earth it is better to think of it as the latter. The two units are of course equivalent, as can be shown by reducing the newton to base units.

If in doubt, it is probably better to use Nkg^{-1} as the units for g. Just to give one example where this is important: imagine carrying out an experiment to measure how g varies from floor to floor of a building. It should reduce with increasing height according to Equation 13.2. It would then be sensible to provide a value for $\frac{dg}{dh}$ — the reduction of g with height. Providing the units for this as s^{-2} would be completely correct but would not convey much information, whereas $Nkg^{-1}m^{-1}$ would provide much more clarity.

The distinction between the differing concepts of g leads neatly to different interpretations of mass.

13.2.3.   *Inertial and Gravitational Mass*

Mass was first defined in Section 2.3, and then an extra layer to the physics was added in Section 9.6. Now we are looking at gravitation, a further definition of mass is necessary:

Inertial mass, m_i can be defined as reluctance for an object to undergo acceleration. It is related to force and acceleration by $m_i = \frac{F}{a}$.

Gravitational mass, m_g can be defined as that quantity which causes an object to experience a force in a gravitational field. It can be expressed in equation format by Newton's law of gravity.

Note that these two definitions are completely different — the only thing linking the two concepts is that they both are related to force, but in very different ways.

After introducing weight in Section 5.2.1, we then said $ma = mg$, but now with these contrasting definitions this must be refined to state $m_i a = m_g g$, and thus an object's acceleration in a gravitational field is given by $a = \frac{m_g}{m_i} g$.

This means that unless $m_g \propto m_i$ then different objects will fall at different rates in a gravitational field. You may well have seen this done before, but try it again now: drop two objects of obviously different mass from the same height at the same time (a hefty set of keys and a pen, for example). They hit the ground at the same time. What you have done is essentially shown is that $m_g \propto m_i$, otherwise they would fall at different rates.

In fact, *experimental evidence suggests* that not only is $m_g \propto m_i$ but that their ratio is unity, i.e.

$$\frac{m_g}{m_i} = 1.$$

This means that inertial mass and gravitational mass can be treated as equivalent quantities. There has never been an experiment done that has shown the two quantities to be different.

Further analysis of this concept can be found in resources on advanced classical mechanics, and also in resources on Einstein's theory of general relativity, which relies on the idea that gravitation can be considered a form of inertia. A full study of this is beyond the scope of this course, but is addressed in many undergraduate degree courses and any postgraduate course on gravitation.

13.3. Gravitational Potential and Binding Energy

The *gravitational potential* V at a point at distance r from an object of mass m is defined as the *work done per unit mass* in bringing another object from infinity to the point. For a point mass, the equation for potential is given by:

$$V = -G\frac{m}{r}. \tag{13.3}$$

Gravitational potential is a scalar with units of Jkg^{-1}.

The potential at the surface of the Earth is therefore $-6.6M\,\mathrm{Jkg}^{-1}$. This means that if an object of 1 kg of mass is brought from a large distance away to the surface of the Earth then it will need $-6.6M\mathrm{J}$ of work to be done *on* it. In other words, $+6.6M\mathrm{J}$ of work will be gained by the object when it moves towards the Earth — in the absence of the friction, the work-energy theorem means that that this will be entirely kinetic energy.

13.3.1.   *Proof of Equation 13.3*

Consider a test mass M that is initially a distance r along the x-axis from a mass m that is fixed in position. Let us calculate how much work is needed to push the test mass *away* to an infinite distance away from m.

The test mass is pushed away at a constant speed so its kinetic energy does not increase — the gravitational force of attraction by the mass m is thus exactly balanced by the contact force supplied by the agent pushing the test mass away, as shown in Figure 13.2.

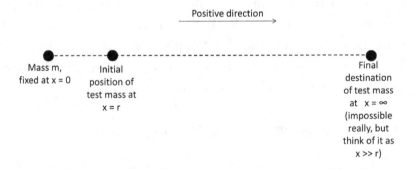

Figure 13.2:   A test mass being pushed a long distance away from r.

The work done in moving the test mass a small displacement δx is given by $\delta W = F \cdot \delta x$ where F is the force on the test mass and δx the infinitesimal displacement that it moves. When a distance x away from the test mass, this force will be $+\frac{GMm}{x^2}$ as it acts left to right (positive in the system sketched in Figure 13.2).

This gives $\delta W = \left(+\frac{GMm}{x^2}\right) \cdot (\delta x) = +\frac{GMm}{x^2} \cdot \delta x$ which is a positive energy, meaning the work done by the contact force is positive — i.e. the total energy of the system is increasing. As the kinetic energy is constant, this means that the potential energy is also increasing as should be expected.

The total work done on moving from r to infinity is thus given by:

$$W = +GmM \int\limits_{x=r}^{x=\infty} \frac{dx}{x^2} = +G\frac{Mm}{r}.$$

This is the known as the *binding energy* of two masses separated by a distance r and is the amount of (positive) energy required to separate them far enough so that their mutual gravitational force of attraction is negligible.

The *potential energy*, U, of the two objects is given by the negative of the binding energy — it is the amount of work required to bring the two objects from infinite separation to a separation r.

To state this generally for two masses m_1 and m_2, the gravitational potential energy is given by:

$$U = -G\frac{m_1 m_2}{r}. \tag{13.4}$$

The potential energy *per unit mass* of an object is hence given by Equation 13.3.

Notice that the gravitational potential of an object is always negative as the zero of potential is *defined* as being at infinity.

Also note that the potential energy belongs to neither one object nor the other, rather it is shared between the two objects (or can be said to be a property of the system) and is also always negative by convention.

It is worth reiterating that the absolute values of gravitational potential and potential energies are meaningless as they have been defined relative to some arbitrary reference point; rather it is the *change* in these quantities from one moment to another that is important.

Though the proof has been performed for a movement along the x-axis, it can be proved that the Equations 13.3 and 13.4 are path independent because the gravitational field is conservative.

13.3.2. *Escape Velocity*

The escape velocity of an object is the initial speed a particle needs to completely escape the gravitational attraction of the object. In essence, this means the velocity required to move to an infinite distance away. Computing the escape velocity is simple using energy arguments.

If the kinetic energy of the particle, mass m, on leaving the surface (distance r from the centre) of an object, mass M, is *just* enough to escape the gravitational field then provided no energy is lost to friction the kinetic energy on launch will equal the potential energy gained so that $\frac{1}{2}mv_{escape}^2 = G\frac{Mm}{r}$.

This leads to the equation:

$$v_{escape} = \sqrt{2\frac{GM}{r}}.$$

Note that the escape velocity is:

(i) Independent of direction of projection and,

(ii) A factor $\sqrt{2}$ greater than the orbital speed needed for an object launched at the same distance from an object's centre.

The escape velocity of the Earth is approximately $11\,k\mathrm{ms}^{-1}$. The escape velocity of an asteroid $1\,\mathrm{km}$ across would be about $1\,\mathrm{ms}^{-1}$ $\left(\text{the field strength at the surface would be about } \frac{1}{5000}g\right)$.

13.3.3. *Black Holes and the Schwarzschild Radius*

A black hole is an object whose gravitational field is so great that not even light can escape its grasp. In other words, the escape velocity is greater than the speed of light. The radius of such an object is called the Schwarzschild radius. It is sometimes naively stated that setting the escape velocity to the speed of light can work in the above calculation. It does, but only by chance. The kinetic energy of light is not given by $\frac{1}{2}mc^2$ and the gravitational potential near a black hole is not $-G\frac{m}{r}$. General relativity is needed to fully work through the problem. Try working through the calculation to see what size the Earth would need to be squashed down to so it becomes a black hole — you should get a diameter of about 1.8 cm.

13.4. Gravitational Effects of A Spherical Shell

The gravitational effects of a thin, spherical shell of material are worth considering. There are two important cases to consider:

(i) The gravitational force on a mass *inside* a hollow sphere and,
(ii) The gravitational force on a mass *outside* a hollow sphere.

Though the final solutions to the problems stated above are fairly easy to take on board and their consequences simple (and fun) to understand, no formal proof of solutions will be given in this textbook. The reasons for this are outlined below, with reference to sources where three different proofs can be found.

By far the most elegant method of addressing the problem uses a principle known as *Gauss's law*. This law applies to any inverse square law field, and is most commonly used in electrostatics but applies equally well to gravity. Unfortunately, the mathematics required is just a little beyond the scope of this textbook (it uses the notion of surface integrals) so it cannot sensibly be included here. The mathematical technique itself is found in first year courses of university-level physics and Gauss's law is usually dealt with in introductory electromagnetism courses.

The most direct way of calculating the force is by the use of integral calculus. This method is rather fiddly, time consuming and inelegant. Although the method is perfectly understandable with the assumed level of mathematics required for readers of this book it is omitted as inclusion is unlikely to aid the understanding of the physics. One example of the proof using this method can be found in *University Physics* by Young and Freedman (2006).

A third method is to calculate the potential at a point in the vicinity of the hollow sphere and use $\mathbf{F} = -\nabla U$ to compute the force. This also requires a doable level of integral calculus and is a little less fiddly than the direct force calculation as it does not involve splitting forces into components, however it is still too cumbersome to merit inclusion. A proof by this method can be found in *Newtonian Mechanics* by A.P. French (1971).

Let us now consider the results with qualitative suggestions as to their validity.

13.4.1.   *The Force on a Mass Outside a Hollow Sphere*

Consider an object outside a hollow sphere of uniform thickness and density, as shown in Figure 13.3.

Consider the force on the object outside the sphere.

It should be clear that there will be no net force on the object in the y-direction: the mass distribution "above" and "below" the object as shown in the figure is perfectly symmetrical, hence the contributions balance. Similarly, there will be no net force in the (unlabelled) z-direction for the same reason.

The net force in the x-direction is less obvious. The neat result that can follow as a result of some rather convoluted geometry and

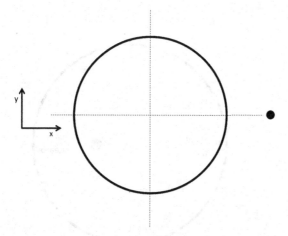

Figure 13.3: **Mass outside a hollow sphere shown as a cross section through the *xy* plane.**

calculus is that the force acts as if *all* the mass comprising the sphere is concentrated at its centre. In other words:

> **The gravitational force on an object outside a uniform spherical shell is equivalent to that of a point mass located at the centre of the shell.**

This is a surprising and important result. It has no dependence on the radius of the sphere and is a unique consequence of the inverse square law.

The effect of the force due to a solid planetary sphere should now be apparent — a sphere can be thought of as having onion-like layers. If each of these layers has a uniform density, each will contribute additively to the effect of the effective mass being at the centre of the sphere, even if the density is a function of depth.

This is an amazing result as it is far from obvious that an object standing on the surface of the Earth would be attracted as though the entire mass were concentrated at the centre $6,000\,km$ beneath. This justifies the treatment of spherically symmetrical objects as point masses when considering their gravitational effect.

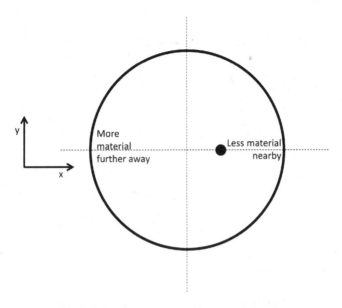

Figure 13.4: **Mass inside a hollow sphere shown as a cross section through the xy plane.**

13.4.2. *The Force on a Mass Inside a Hollow Sphere*

Consider an object inside a hollow sphere of uniform thickness and density, as shown in Figure 13.4.

Consider the forces on the object inside the sphere. As the sphere completely surrounds the object there will be a gravitational force acting on it in all directions (with just the x- and y-directions shown in the cross section).

It should be clear that there will be no net force on the object in the y-direction: the mass distribution "above" and "below" the object as shown in the figure is perfectly symmetrical thus the contributions balance. Similarly, there will be no net force in the (unlabelled) z-direction for the same reason.

The net force in the x-direction is, as before, less obvious. As indicated on the figure, there is less material on the right-hand side of the object than on the left, but the material on the right-hand side is closer to the object than the material on the left. So which side wins — the left-hand side with more material a greater distance away, or the right-hand side with less material but a closer distance to the object?

The lovely resolution to this is that the two contributions are equal and opposite; the extra force generated by the excess of mass on the left is exactly compensated for by the extra distance. This is true regardless of the position of the object. In other words:

> **The gravitational force on an object inside a uniform spherical shell is *zero at all points inside the shell*.**

This is a case where, as there is zero gravitational force on the body inside the sphere, the body would be genuinely weightless. So if there were such a thing as a hollow planet, if one tunnelled inside the planet one would float freely inside.

As well as provoking amusing thought experiments and science fiction ideas, this does have real importance. If tunnelling into the Earth, the gravitational field of all outer shells will have zero effect, but the inner ones act as if the mass is in the centre. The gravitational field of a uniformly dense sphere has the form described in Figure 13.5.

Therefore, if a person tunnelled all the way to the centre of the Earth then they would experience no gravitational force due to the Earth and would be genuinely weightless. They would, of course, feel the effects of gravity in a different way as the normal contact force due to the combined gravitational effects of the surrounding mass manifesting itself as a summed effect from the surface to the centre, leading to an enormous pressure (of the order $300\,GPa$); aside from the high temperatures and hostile chemical environment they would not be at all comfortable. Science fiction writers may occasionally

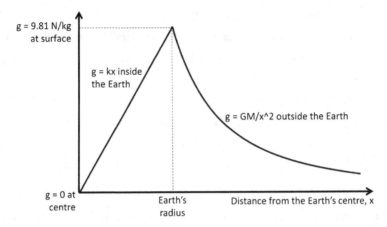

Figure 13.5: **Variation of field strength with depth for a planet Earth of uniform density.**

state that the centre of a planet is an oasis of calm because the field is zero, but in reality this would not be the case.

13.5. Planetary Variations in Field Strength

Density variations in the Earth mean the field increases slightly for the first few hundred kilometres of depth before falling almost uniformly.

Gravitational field strength on the Earth's surface varies in magnitude (and direction) from about $9.83 \, \text{ms}^{-2}$ (at the poles) to $9.79 \, \text{ms}^{-2}$ (at the Equator) due to:

- The flattening of the Earth — the planet is "oblate spheroid" in shape rather than a perfect sphere.
- Altitude — the higher you are, the further you are from the centre of the Earth.
- Local density variations — up to $10^{-5} \, \text{Nkg}^{-1}$ when moving across areas of differing geology.

- The rotation of the Earth — the spinning causes a smaller *apparent* weight at the equator (how much can be worked out using the physics of motion in a circle).
- Tides — sensitive gravity meters can detect periodic changes.

All of these effects are measurable and quantifiable, and put to good use. Geophysical surveys, for example, observe the changes in field strength over areas of land to determine the nature of the underlying rock type.

14

Rotational Analogues

This chapter introduces the defining quantities and equations associated with the rotation of rigid bodies. Many of the results are directly analogous to the linear relationships introduced in Chapters 2–4. The main added complication with rotational quantities is the conceptual difficulty of treating them as vectors with the mathematical corollary of the use of cross products in equation manipulation. This chapter serves as a brief introduction; details and examples with respect to the physics follow in Chapters 15–19.

14.1. Angular Velocity

Consider a particle moving in plane at constant speed v at fixed distance r from the origin, O, as shown in Figure 14.1.

The *angular velocity*, $\boldsymbol{\omega}$, of the particle is *defined* as the rate of change of angle, θ, with time, i.e.

$$\boldsymbol{\omega} = \frac{d\theta}{dt}. \tag{14.1}$$

It is a *vector* with the direction given by the right-hand grip rule: make a thumbs up sign with your right hand with the fingers curling in the same direction that the particle moves in. The pointing thumb is in the direction of the angular velocity vector. This is out of the plane of the paper in the case of Figure 14.1. Angular velocity has SI units of rad.s^{-1}. For example, a record player's turntable revolving at $33\frac{1}{3}$ revolutions per minute has an angular velocity of approximately 3.5 rad.s^{-1} out of the plane.

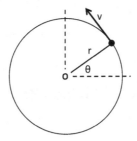

Figure 14.1:   A particle moving in a circle at constant speed about the origin.

In a small time increment, the small angle turned by the particle is given by $\delta\theta = \frac{\delta x}{r}$.

Taking the time derivative of both sides of this equation gives $\frac{\delta\theta}{\delta t} = \frac{1}{r} \cdot \frac{\delta x}{\delta t}$, hence $\omega = \frac{v}{r}$ from the definitions of linear and angular velocity. Incorporating the vector notation,

$$\boldsymbol{v} = \boldsymbol{\omega} \times \boldsymbol{r}. \tag{14.2}$$

If you are not happy with angular velocity being classed as a vector in this way then that is all well and good. Further discussion on the concept is provided in Section 14.6.

14.2.   Angular Acceleration

If the particle in Figure 14.1 changes its speed but keeps its distance from the origin constant then it undergoes an *angular acceleration*, α, given by:

$$\alpha = \frac{d\omega}{dt} = \frac{d^2\theta}{dt^2}. \tag{14.3}$$

Angular acceleration is a vector quantity with its direction in the direction of the instantaneous *change* in angular velocity (if speeding up, in the same direction; if slowing down, in the opposite direction). It has SI units of rad·s^{-2}.

On switching on a turntable the angular acceleration will be approximately 2 rad·s^{-2} in the same direction as the angular velocity.

Angular acceleration is related to the instantaneous linear acceleration of the particle by:

$$a = \frac{dv}{dt} = \frac{d\omega}{dt} \times r = \alpha \times r. \qquad (14.4)$$

It is possible at this stage to work through a series of rotational kinematic equations as presented in Section 3.1 for linear motion. This is an exercise of limited use for this textbook but may be useful practice for the reader.

14.3. Rotational Kinetic Energy and Moment of Inertia

14.3.1. *Single Particle*

The kinetic energy of the particle in Figure 14.1 is given by $\frac{1}{2}mv^2 = \frac{1}{2}mr^2\omega^2$.

The *rotational kinetic energy* E_{KR} of the particle can therefore be written:

$$E_{KR} = \frac{1}{2}I\omega^2, \qquad (14.5)$$

where

$$I = mr^2 \qquad (14.6)$$

is the *moment of inertia of the particle about the origin.*

The moment of inertia of the particle is a scalar with units of kg·m². A particle of 1 kg situated at 1 m from an axis of rotation will have a moment of inertia of 1 kg·m². Note that this value is irrespective of motion — the moment of inertia relative to the origin is a quantity that the system has whether there is rotation or not, and that to sensibly define the moment of inertia the axis of rotation must be specified.

14.3.2. *Several Particles*

Now consider a solid object made of N particles rotating about the origin, as shown in Figure 14.2.

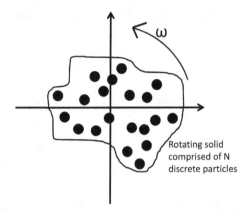

Figure 14.2:   A solid rotating object made of many discrete particles.

In this case, the total rotational kinetic energy will be given by:

$$\sum_{i=1}^{N} \left(\frac{1}{2}m_i v_i^2\right) = \frac{1}{2}\sum_{i=1}^{N}(m_i r_i^2 \omega^2) = \frac{1}{2}\omega^2 \sum_{i=1}^{N}(m_i r_i^2) = \frac{1}{2}I\omega^2,$$

where the moment of the inertia of the system of particles is given by:

$$I = \sum_{i=1}^{N} m_i r_i^2. \tag{14.7}$$

14.3.3.   *Continuum of Particles*

If the rotating object is considered as a continuous distribution rather than a system of particles then the rotational kinetic energy will still be $\frac{1}{2}I\omega^2$, with the moment of inertia given by an integral over the distribution rather than a sum. If the density of the object is *uniform,* this can be written:

$$I = \int r^2 dm, \tag{14.8}$$

where the limits of the integral encompass the volume of the rotating object.

14.3.4. *Meaning of Moment of Inertia*

The formulae for kinetic energy clearly show how moment of inertia can be used as a direct analogy for mass when comparing linear and rotational motion. This can be taken further: just as mass can be defined as a reluctance to undergo a *linear* acceleration (rate of change of linear velocity), moment of inertia can alternatively be defined as a reluctance to undergo a *rotational* acceleration (rate of change of angular velocity). In other words, it is a measure of how difficult it is to rotate an object (or stop it from rotating).

Note that the moment of inertia of an object depends on the axis of rotation and the position of the origin — a value will be meaningless unless these are stated.

For objects of *uniform density*, the formula for moment of inertia is always given by $I = kmr^2$ where k is a dimensionless constant and r is the distance from the origin to the centre of mass.

14.3.5. *Common Examples*

Two worked examples are given here, followed by several well-known examples. The worked examples concern a uniform rod. Before going through them, try the following experiment: take a uniform rod of some kind, the bigger the better (a broom handle is ideal). Hold the rod with one hand in the middle and rotate it by twisting your wrist. Now hold the rod near the end and try and rotate it again by twisting your wrist. You will find that it is much harder to do than before. You have now qualitatively demonstrated that there is more resistance to an angular acceleration at the end of a rod than at its centre — i.e. the moment of inertia is greater about the end than the centre. The following examples prove and quantify this.

14.3.5.1. *The moment of inertia of a thin rod about its centre*

Consider a thin uniform rod of mass M and length L being rotated about its centre, as illustrated in Figure 14.3.

Consider the moment of inertia, δI, of the shaded element of thickness δx and mass $\frac{\delta x}{L} M$:

From the definition of moment of inertia, $\delta I = \frac{\delta x}{L} M x^2$. The moment of inertia of the whole rod will therefore be:

$$I = \int_{x=-\frac{L}{2}}^{x=\frac{L}{2}} \frac{M}{L} x^2 \, dx = \frac{M}{L} \int_{x=-\frac{L}{2}}^{x=\frac{L}{2}} x^2 dx.$$

This is evaluated to give:

$$I = \frac{1}{12} ML^2 \text{ for a rod rotating about its centre.}$$

14.3.5.2. *The moment of inertia of a thin rod about its end*

Consider the same thin uniform rod of mass M and length L being rotated about its end rather than its centre; the coordinates are changed to suit the new set-up, as shown in Figure 14.4.

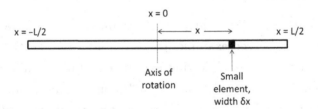

Figure 14.3: A uniform rod with an axis of rotation through the centre.

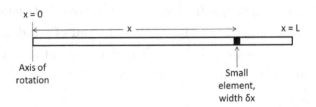

Figure 14.4: A uniform rod with an axis of rotation through the end.

Repeating the procedure from the previous example:

Consider the moment of inertia, δI, of the shaded element of thickness δx and mass $\frac{\delta x}{L}M$:

From the definition of moment of inertia, $\delta I = \frac{\delta x}{L}Mx^2$. The moment of inertia of the whole rod will therefore be:

$$I = \int\limits_{x=0}^{x=L} \frac{M}{L}x^2 dx = \frac{M}{L} \int\limits_{x=0}^{x=L} x^2 dx.$$

This is evaluated to give:

$$I = \frac{1}{3}ML^2 \text{ for a rod rotating about its end.}$$

The moment of inertia of a rod is thus four times greater about its end than about its centre — you can say it is four times as hard to rotate the rod about the end than about the centre.

Some other well-known moments of inertia are provided without proof (or usually without proof) below.

14.3.5.3. *Hoop or hollow cylinder*

Consider a hoop or a hollow cylinder of mass M and radius R rotating about an axis, as shown in Figure 14.5.

If the hoop or cylinder is infinitesimally thin then all its mass is set at a distance R from the axis of rotation. The moment of inertia will therefore be the same as a point mass set at this distance, i.e.

$$I = MR^2 \text{ for a hoop or hollow cylinder.}$$

14.3.5.4. *Disc or solid cylinder*

If we try and rotate a disc or solid cylinder then not all the mass is concentrated at the edge; thus for a solid cylinder of the same mass and dimension as the hollow cylinder, the moment of inertia should be less. In fact, it turns out to be exactly one half of that of the hollow cylinder, i.e.

$$I = \frac{1}{2}MR^2 \text{ for a disc or solid cylinder.}$$

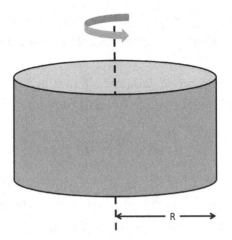

Figure 14.5: A rotating hollow cylinder.

14.3.5.5. *Hollow sphere*

Stating without proof again, for a hollow sphere rotating about its centre:

$$I = \frac{2}{3}MR^2 \text{ for a hollow sphere.}$$

14.3.5.6. *Solid sphere*

This should be less than for a hollow sphere of the same dimension when rotating about its centre, and in fact:

$$I = \frac{2}{5}MR^2 \text{ for a solid sphere.}$$

This is the example that will be used most frequently in Chapter 17 on rolling objects.

14.4. Torque

Consider putting a force on a spanner causing it to turn, as demonstrated in Figure 14.6.

The force causes a turning effect known as the *torque*, Γ. The torque on an object about a point is *defined* as the cross product of

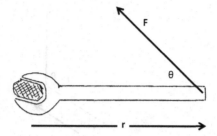

Figure 14.6: **A force causing a turning effect about a pivot.**

the displacement from the pivot to the force with the force itself, i.e.

$$\boldsymbol{\Gamma} = \boldsymbol{r} \times \boldsymbol{F}, \tag{14.9}$$

with a magnitude given by $rF\sin\theta$ and the direction given by the right-hand rule (out of the plane in Figure 14.4).

Torque is a vector with SI units of Nm. Notice that these are the same base units as for energy and work; both are products of a force with a distance but one is a scalar dot product and the other a vector cross product. Of course, work uses the joule as the SI unit.

This direction is given by convention and notice how it works in conjunction with the right-hand grip rule for angular velocity. The end of the spanner has an angular velocity about the head of the spanner when it rotates, in the same direction as the torque.

14.4.1. *Rotational Equivalent of Newton's Second Law*

The linear acceleration of a point on the spanner is given by $\boldsymbol{F} = m\boldsymbol{a} = m\frac{d\omega}{dt} \times \boldsymbol{r}$ (from Equation 14.4). Hence, the torque can be expressed by $\boldsymbol{\Gamma} = \boldsymbol{r} \times m\frac{d\omega}{dt} \times \boldsymbol{r} = mr^2\frac{d\omega}{dt}$ (using the rules for cross products).

The resultant external torque on a particle can thus be written in the form:

$$\boldsymbol{\Gamma} = I\frac{d\omega}{dt}, \tag{14.10}$$

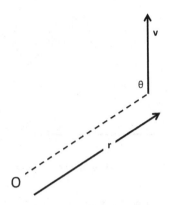

Figure 14.7:   A particle moving relative to a fixed point.

and this can be generalised to any macroscopic object. This is recognisable as the rotational form of Newton's second law, i.e. $\boldsymbol{F} = m\boldsymbol{a}$.

14.5.   Angular Momentum

Consider a particle of mass m moving at velocity $\boldsymbol{v}$ at displacement $\boldsymbol{r}$ from a point O, as illustrated in Figure 14.7.

The *linear* momentum of the particle relative to O is $\boldsymbol{p} = m\boldsymbol{v}$ and the *angular* momentum, $\boldsymbol{L}$, of the particle about the point is *defined* by:

$$\boldsymbol{L} = \boldsymbol{r} \times \boldsymbol{p} = m\boldsymbol{r} \times \boldsymbol{v} \qquad (14.11)$$

and has magnitude $L = mrv\sin\theta$ and direction given by the right-hand rule.

Angular momentum is a vector with SI units of N·m·s or kg·m^2·s^{-1}; both are commonly used. Chapter 18 is devoted to discussing angular momentum in more detail.

14.6.   A Bit More on Scalars, Vectors, and Tensors

In Section 14.1, angular velocity was introduced as a vector with the direction given by the right-hand grip rule. If this was the first time you had encountered this concept then some confusion is entirely to be expected.

Things are a little more complicated than this elementary analysis provides, and a deeper understanding of the concepts will likely not arrive until you have studied mathematics and physics to a higher level — normally until about the third year of a standard university degree. The remaining chapters go into more detail about rotational motion, making use of the definitions provided using examples where the physics can safely be investigated, and pointing out where subtleties creep in that require a more sophisticated analysis.

Nevertheless, two big clues as to the subtleties of rotational motion are provided now to pique your interest, provoke some further reading (there are plenty of decent books out there but the Internet has a huge amount of good quality information too) and allow you to be able to exercise your own judgement as to when the physics provided in this book reaches its limits.

14.6.1. *Angular Velocity vs. Linear Velocity*

Try the following fun exercise, in two parts: sketch a car or bicycle, or any vehicle with wheels, moving along the ground *towards* a mirror. Now sketch the reflection of the vehicle including the vector arrows for the linear velocity of the vehicle, and the linear velocity of the reflection. It is not important if the drawing of the vehicle is not very good, but the vector arrows need to be nice and clear.

Using the right-hand grip rule, sketch the direction of the angular velocity of one of the wheels (by convention you can use a × symbol for into the plane and a ⊙ symbol for out of the plane).

For part two, carry out exactly the same exercise with the car travelling *parallel* to the mirror.

How do the linear velocities reflect in the two cases? How about the angular velocities?

This exercise illustrates a fundamental difference between the familiar vectors of linear motion and the vectors introduced in this chapter. If you felt that there was something suspect about them, then you were indeed right! Angular velocity is in fact an example of a *pseudovector*. There are several other examples in the discipline; other examples can be seen in electromagnetism and fluid dynamics.

14.6.2.   *The Moment of Inertia Tensor*

Moment of inertia has been introduced as a scalar quantity, with the additional (and always correct) message that as well as the point of rotation being specified, there are three components of the quantity depending on what axis the object is rotated around. Really, the moment of inertia of a three-dimensional object is expressed by a 3×3 matrix where the diagonal elements of the matrix are the three moments of inertia previously discussed. For symmetrical objects, the other elements in the matrix are zero, but for asymmetrical objects the other matrix elements are given by a quantity known as the *product of inertia*, which the reader can look up.

When represented as such, the moment of inertia is certainly not a scalar, nor is it a vector, but is a quantity known as a *tensor*. Tensors are sometimes regarded as esoteric in physics, possibly because they are heavily utilised in advanced topics such as general relativity and quantum field theory. In fact, they are ubiquitous, appearing in all branches of physics. Familiar concepts such as Hooke's law and Ohm's law, for example, require the use of tensors when treated properly in three dimensions.

15

Equilibrium and Balance

This chapter introduces the concepts of the centre of gravity, mass and buoyancy, and looks at situations involving equilibrium and balance.

15.1. Centre of Mass

The centre of mass can be defined as follows:

> **The centre of mass of an object is the weighted average location of all its mass.**

15.1.1. Discrete Particle System

For an object made of a *discrete system* of particles each of mass m_i and displacement r_i, the position r_{CM} from the origin O of the centre of mass is given the formula:

$$r_{CM} = \frac{\sum m_i r_i}{M}, \tag{15.1}$$

where $M = \sum m_i$ is the total mass of the object.

For example, consider the system of fixed particles in Figure 15.1.

Taking the origin at (0,0) and using Equation 15.1, the x-coordinate of the centre of mass will be at $\frac{4 \cdot 1 + 1 \cdot 2 + 2 \cdot 4}{7} = 2\,\mathrm{m}$. The y-coordinate will be at $\frac{4 \cdot 4 + 1 \cdot 1 + 2 \cdot 2}{7} = 3\,\mathrm{m}$.

Hence, the location of the centre of mass is $r_{CM} = (2i + 3j)\,m$ (shown by a star in the figure).

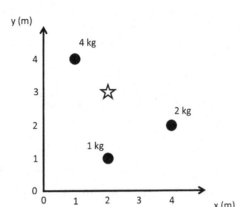

Figure 15.1:   A system of three fixed particles.

15.1.2.   *Continuum System*

For an object made of a *continuum* of particles, the analogous equation to Equation 15.1 for objects of *uniform density* is:

$$r_{CM} = \frac{1}{M} \int r \cdot dm. \tag{15.2}$$

It is usually convenient to split the equation into components, i.e.

$$x_{CM} = \frac{1}{M} \int x \cdot dm,$$

$$y_{CM} = \frac{1}{M} \int y \cdot dm,$$

$$z_{CM} = \frac{1}{M} \int z \cdot dm.$$

For example, consider a right-angled triangle of mass M, such as in Figure 15.2.

The mass of the narrow element in the figure is given by $\delta m = 2\frac{x \cdot \delta x}{a^2} M$ (work out why for yourself). Hence, the x-coordinate of the centre of mass is given by:

$$x_{CM} = \frac{2}{a^2} \int\limits_{x=0}^{x=a} x'^2 dx' = \frac{2}{3}a,$$

i.e. two-thirds of the way in from the vertex.

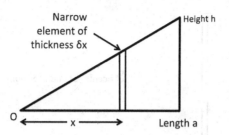

Figure 15.2: A uniform right-angled triangle.

By symmetry, the vertical coordinate of the centre of mass will lie two-thirds of the way in from the top.

15.1.3. *L-Shaped Object*

If an object has a fixed shape then the centre of mass will be at a fixed position relative to the shape. If an object can change its shape then the centre of mass can change position. This can be shown by considering a stick which can be bent to form an L, as shown in Figure 15.3.

This image illustrates an important point — the centre of mass of an object does not necessarily lie inside the object.

15.1.4. *Importance*

When a resultant external force acts on an object then the object accelerates according to $a = \frac{F}{m}$. This is fine for a point particle but what about for a complicated object (like a human) with flailing limbs and a bobbing head? It cannot very well be said that all parts accelerate with magnitude a; in fact, it is always the *centre of mass* that undergoes this acceleration. This is easy enough to take on board but a little trickier to prove; that it must be so will become clearer when angular momentum is covered in more detail.

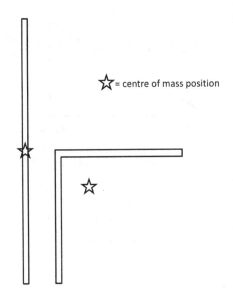

Figure 15.3: Centre of mass for an *L* shape.

15.2. Centre of Gravity

The centre of gravity of a body is the point about which the net gravitational torque is zero. It is the mean location of the gravitational force on a body. The total *weight* of a body can be thought to originate from its centre of gravity. If an object is in a *uniform* gravitational field, the centre of mass and the centre of gravity will be at exactly the same spot.

If an object is in a non-uniform gravitational field then the centre of mass and centre of gravity can and will be in different places. For example, consider a uniform pole standing on the surface of the Earth, tilted at some angle to the upright (so the pole does not point directly towards the Earth's centre). The centre of mass of the uniform pole is exactly at its centre. But as the lower end of the pole is relatively close to the Earth and the higher end of the pole is farther away, the torque on the lower half will be greater. The position of zero net torque is therefore lower down the pole than the centre of mass. This effect will only be significant for a particularly long pole, of course — one that has a length comparable to the radius of the Earth.

15.3. Centre of Buoyancy

If an object has an upthrust (buoyancy) force on it then the force acts from the centre of buoyancy — this is the centre of gravity of the *displaced fluid*.

Consider a uniform stick with a lower density than water placed in the water at an angle θ, as demonstrated in Figure 15.4.

In this situation, the weight acts downwards from the centre of gravity of the stick, which is approximately halfway along its length, and the upthrust acts upwards from the centre of gravity of the displaced fluid, which is a little further along. The two forces act to rotate the stick anticlockwise as shown on the diagram, which explains why a pencil floats "on its side".

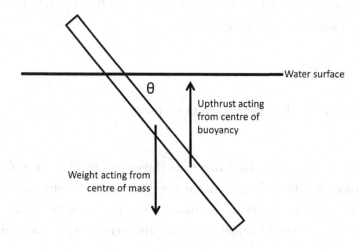

Figure 15.4: A stick in water.

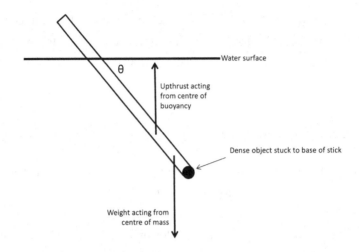

Figure 15.5: A stick in water with plasticine on the end.

To get a stick to float upright, a heavy object (like a piece of plasticine) can be placed on the end; provided the weight can still be matched by the upthrust, this will work (see Figure 15.5).

In this situation, the centre of gravity is shifted significantly downwards, but the centre of buoyancy less so. If the centre of gravity is now below the centre of buoyancy (as in the figure) then the stick rotates clockwise to float upright.

15.4. Equilibrium

For an object to be in equilibrium then the following conditions must hold:

(1) The resultant external **force** on the object must be zero (or the object would undergo a linear acceleration according to $a = \frac{F}{m}$).

(2) The resultant external **torque** on the object must be zero (or the object would undergo an angular acceleration according to $\frac{d\omega}{dt} = \frac{\Gamma}{I}$).

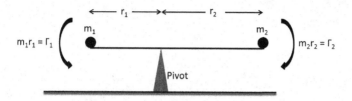

Figure 15.6: Masses on a see-saw.

Hence, if an object is seen to be in equilibrium, it can be concluded that the forces and torques must add to zero. Conversely, if forces and torques add to zero, it can be concluded that the object is in equilibrium.

15.5. Examples of Equilibrium

15.5.1. *See-Saw*

Consider two masses on a light see-saw balanced at a pivot point, as shown in Figure 15.6.

For the system to be in equilibrium, the torques must balance so $m_1gr_1 = m_2gr_2$. It is thus required that $\frac{r_1}{r_2} = \frac{m_2}{m_1}$. This is the equivalent of saying that the see-saw balances at the centre of gravity/mass.

15.5.2. *Balancing Pencil*

If the weight force remains exactly above the pivot point, the pencil is in equilibrium. However, if the pivot is indeed a *point*, it is impossible for the weight to be perfectly aligned and the pencil is in *unstable equilibrium* (see Figure 15.7).

Of course, the pencil will topple quite quickly (the mechanics of toppling is investigated in the next chapter). The pencil can only stand upright forever if the weight is *exactly* above the contact force and the pencil is perfectly still. Incidentally, one classic problem in physics is the question of how long it is physically possible to have the

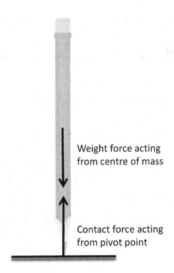

Weight force acting
from centre of mass

Contact force acting
from pivot point

Figure 15.7:   A pencil balancing on its end.

pencil stay upright. This will not be discussed fully here as the answer
requires quantum mechanics — particularly Heisenberg's uncertainty
principle, which states that an object cannot be both perfectly still
and have its position uniquely determined. You are encouraged to
try to make this estimation for yourself once you have covered the
topic properly, but the answer is that it is impossible to balance the
pencil for longer than about 4 s.

15.5.3.   *Leaning Ladder*

Consider a person of mass m climbing a light ladder leaning against
a *smooth* wall and *rough* floor. The free body diagram for the ladder
with the person represented as a black dot, which also incorporates
the important distances, is displayed in Figure 15.8.

For what value of x is the ladder most likely to slip?

If the ladder is in equilibrium then:

(1) The vertical forces must balance so $N_1 = mg$.
(2) The horizontal forces must balance so $F = N_2$.

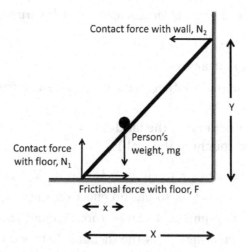

Figure 15.8: **Free body diagram for the ladder.**

(3) The torques must balance. Torques can be taken *about any point* and give the same result so if we *choose* to take them about the contact point with the floor then $mgx \, (clockwise) = N_2Y \, (anticlockwise)$.

Note that neither N_1 nor F contribute to the torque about the contact point with the floor as they act through this point and thus the displacement in $\mathbf{\Gamma} = \mathbf{r} \times \boldsymbol{force}$ is zero.

This gives $mgx = FY$.

Hence, the frictional force keeping the ladder from slipping is given by $F = \left(\frac{mg}{Y}\right) x$.

The distance, x, along the ground between the foot of the ladder and the point directly underneath the person is the only variable in this equation. This means that the frictional force is zero at the base and gets greater the further the person moves up the ladder — i.e. it becomes more likely that the person will slip (an intuitive result for anyone who has ever climbed a ladder).

15.5.3.1.   *Check: taking the torques about an arbitrary point*

It may have seemed a little too easy to simply state that the torques can be taken about any point and go from there. Let us check that taking torques about another point gives the same result:

Choosing the other end of the ladder
(i.e. the top end touching the wall)

In this case, the contact force with the wall, N_2, acts through the point under consideration so has no effect on the torque.

The weight, mg, puts a clockwise torque about the point of magnitude $mg(X - x)$ where X is the distance between the base of the ladder and the wall.

The contact force with the floor, N_1, puts an anticlockwise torque about the point of magnitude $N_1 X$.

The frictional force puts a clockwise torque about the point of magnitude FY.

So for equilibrium, $\Gamma_{clockwise} = \Gamma_{anticlockwise}$, i.e. $mg(X - x) + FY = N_1 X = mgX$ as $N_1 = mg$. The mgX contributions cancel giving $FY = mgx$, which is the same result as before.

Choosing the intersection point where the wall meets the floor

This time the frictional force acts through the point under consideration so makes no contribution to the torque.

The weight puts a clockwise torque of $mg(X-x)$ about this point.

The contact force with the wall puts a clockwise torque of $N_2 Y$ about this point.

The contact force with the floor puts an anticlockwise torque of $N_1 X$ about this point.

As before, for equilibrium, $\Gamma_{clockwise} = \Gamma_{anticlockwise}$, so $mg(X - x) + N_2 Y = N_1 X$, which is equivalent to $mgX - mgx + FY = mgX$. Once again, this simplifies to give the same answer as before.

The reader can try other points (try a completely arbitrary point well away from the ladder to be completely general). The beauty of being able to choose is that the equations can be simplified by choosing a point where as many contributions as possible are zero. The initial choice meant that two of the forces did not contribute, giving the easiest route to the solution.

15.5.3.2. *Ladder variants*

This is just one example of the ladder problem in mechanics. There are several others of different complexity. Consider, for example, a wall that is not smooth, a ladder that is not massless, a flexible ladder, one ladder propped up against another or a smooth floor. Note that in the smooth floor case the ladder is *never* stable: the only force that can ever balance the upper horizontal contact force by the wall is the friction. If this is zero, the ladder always just slides down the wall. To picture this, consider a ladder resting against a wall on hard ground with the base placed on a skateboard. No matter how great the frictional force between the ladder and the wall, the ladder slides. (Or consider yourself as the ladder and put your feet on the skateboard and staying as straight as possible with arms outstretched lean against the wall for the same effect.)

16

Unbalanced Objects

This chapter studies two symmetrical systems in which the resultant external torque causes an angular acceleration on the system. No new fundamental physics is introduced and both systems are very familiar but are now analysed with knowledge of the equations of rotational motion. Magnitudes of quantities will usually be used in this section so vector notation will be omitted. The whole chapter comprises two examples, one short and one long. The second takes the physics quite far in order to show how far we can now take the analysis of a simple, though deep, everyday problem.

16.1. An Unbalanced Light See-Saw

Consider a light see-saw pivoted in the centre with unequal masses on each end (see Figure 16.1). Find the initial angular and linear acceleration of the ends of the see-saw.

The *clockwise* torque *about the pivot* (using Equation 14.9) is $\frac{m_2 L g}{2} - \frac{m_1 L g}{2} = \frac{1}{2} L g (m_2 - m_1)$.

The moment of inertia *about the pivot* (see Section 14.3) is $m_2 \left(\frac{L}{2}\right)^2 + m_1 \left(\frac{L}{2}\right)^2 = \frac{1}{4} L^2 (m_2 + m_1)$.

Hence, using $\Gamma = I \frac{d\omega}{dt}$ gives:

$$\frac{d\omega}{dt} = \frac{\frac{Lg}{2}(m_2 - m_1)}{\frac{L^2}{4}(m_2 + m_1)} = 2\frac{g}{L}\left(\frac{m_2 - m_1}{m_2 + m_1}\right).$$

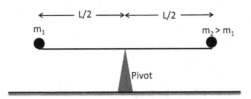

Figure 16.1:   An unbalanced see-saw.

This shows that the angular acceleration is directly proportional to the field strength and inversely proportional to the length of the see-saw.

The tangential linear acceleration is given by $\frac{d\boldsymbol{\omega}}{dt} \times \boldsymbol{r}$ and has magnitude $g\left(\frac{m_2-m_1}{m_2+m_1}\right)$ at the ends of the see-saw.

Note that when the masses are equal, the linear acceleration is zero and when the one mass is much larger than the other the acceleration becomes g as should be expected.

Also, note that this is the *initial* acceleration of the see-saw only; the value changes once the see-saw is no longer parallel to the ground.

16.2.   Rigid Object Toppling About A Pivot

Consider an object of mass m pivoted freely at one end and allowed to fall under gravity. The base of the object is fixed in position and can rotate freely about this base point. This could be either an object with a pointed end toppling on a rough table of arbitrarily large friction, or a hinged object, or an object with a hole at the base with a frictionless nail through it. Figure 16.2 represents the object as a thin stick and shows it as a free body diagram.

16.2.1.   *The Forces*

There are always three forces acting on the object when it topples:

- The weight of the object acts straight downwards from the centre of mass. This is constant.

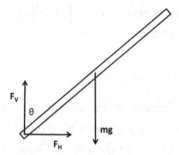

Figure 16.2: Free body diagram for a toppling object (in this case a thin rod) pivoted at one end.

- The horizontal force at the pivot F_H varies with the angle θ. If an object topples on a table, this will be the frictional force.
- The vertical force at the pivot F_V also varies with the angle. If an object topples on a table, this will be the normal contact force.

16.2.2. *Unstable Equilibrium*

If the objects starts from vertical with the centre of mass above the pivot ($\theta = 0$) then it will be in unstable equilibrium. This is analogous to the examples in Sections 11.2 and 15.5.

In this case, $mg = F_V$ and $F_H = 0$.

As there is zero torque about the pivot there will be no angular or linear acceleration, but the slightest push or deviation will cause it to fall.

16.2.3. *Stable Equilibrium*

If the objects starts from vertical with the centre of mass below the pivot ($\theta = \pi$, so it is hanging) then it will be in stable equilibrium.

As with unstable equilibrium, $mg = F_V$ and $F_H = 0$.

As there is zero torque about the pivot, there will be no angular or linear acceleration. Pushes or deviations cause it to move back to the equilibrium point and oscillate. The physics of the oscillation is studied later in this section.

16.2.4.   *Toppling*

If the object starts from any displacement $\theta > 0$, then the torque *about the pivot* is given by $mgl \sin \theta$ where l is the *distance from the pivot to the centre of mass of the object*. From the rotational version of Newton's second law, $mgl \sin \theta = I \frac{d\omega}{dt}$ where I is the object's moment of inertia *about the pivot*.

This gives an angular acceleration as a function of angle of:

$$\frac{d\omega}{dt} = \frac{d^2\theta}{dt^2} = \frac{mgl}{I} \sin \theta.$$

This is a *second-order, non-linear, ordinary, homogeneous differential equation*. It has no general solution (try it on Wolfram Alpha, for example — it is almost identical to the differential equation obtained in Section 12.3 on SHM before applying the small angle approximation) but corroborates common sense that the object will have zero angular acceleration when vertical ($\sin 0 = 0$ for the unstable equilibrium case and $\sin \pi = 0$ for the stable equilibrium case) and a maximum (of $\frac{mgl}{I}$) when horizontal.

16.2.5.   *Accelerations for a Uniform Rod (with a Note on Why Balancing a Pencil on Your Fingertip is Difficult But Balancing a Broom Handle is Easy)*

Up until this point, the analysis in question has been completely general regarding the toppling object. To simplify matters, henceforth the shape will be specified to be a uniform rod. Any other object with a known formula for the moment of inertia could be used instead. The rod is not necessarily the simplest object to consider either — you may want to work the example through with another toppling solid to see how the results compare.

If the toppling object is a *uniform* rod of length L then the moment of inertia *about the pivot* is $I = \frac{1}{3}mL^2$ with a distance to the centre of mass of $l = \frac{1}{2}L$.

Applying these values to the general formula gives an angular acceleration for the rod of:

$$\frac{d\omega}{dt} = \frac{mg\left(\frac{1}{2}L\right)}{\frac{1}{3}mL^2}\sin\theta = \frac{3}{2}\frac{g}{L}\sin\theta.$$

Notice how the angular acceleration is *inversely proportional* to length. We now inspect the middle term in the equation.

The numerator in the equation — $mg\left(\frac{1}{2}L\right)$ — is the torque about the pivot. The torque is the turning effect of the weight, and the longer the rod, the greater this value will be. In fact, they are directly proportional, thus the longer the rod, the greater the effect in pulling it downwards becomes.

The denominator in the equation — $\frac{1}{3}mL^2$ — is the moment of inertia about the pivot, i.e. the reluctance to turn. In this case, the value also increases with length but is proportional to the square of the value. So the longer the rod, the more resistant it is to turning.

As the reluctance to turn increases with length more rapidly than the turning effect, the net effect is that the longer the rod the less readily it turns, despite the torque being greater. This is manifest in the equations as the $\frac{L}{L^2}$ terms cancel to give $\frac{d\omega}{dt} \propto \frac{1}{L}$.

This explains why trying to balance a pencil on one's fingertip is very difficult, but doing the same with a much longer object like a broom handle is much easier — the toppling process is more sluggish so it is easier to correct by moving your finger so as the handle remains upright.

Through a small extension of this analysis, this also explains why balancing a non-uniform rod with the centre of mass closer to one end than the other is easiest with the centre of mass being higher up. Try this with a snooker cue: balancing using the small end is much easier than balancing using the broad end.

16.2.6.   *The Tangential Linear Acceleration and a Surprising Result*

As $\frac{d\omega}{dt} = \frac{3}{2}\frac{g}{L}\sin\theta$ then using Equation 14.4 the tangential linear acceleration at a point at distance r along the rod's axis is given by $a_T = \frac{3}{2}\frac{g}{L}r\sin\theta$.

If the rod is in the horizontal position then this will be $a_T = \frac{3}{2}\frac{g}{L}r$. This equation tells us the downward acceleration at any point along the pivoted rod's axis when it is parallel to the ground. This little result may look innocuous and maybe not of much importance but does lead to an interesting result.

At the centre of the rod, $r = \frac{1}{2}L$ and the tangential acceleration downwards is $\frac{3}{4}g$.

At a position two-thirds along the rod, so $r = \frac{2}{3}L$, the tangential acceleration downwards will be g exactly. Beyond this point the linear downward acceleration is actually greater than g.

At the extreme end of the rod where $r = L$, this gives the surprising result that the tangential acceleration downwards at the end when horizontal is $\frac{3}{2}g$.

This is unusual. Most situations we meet in classical mechanics (e.g. inclined planes, pulley systems and levers) tend to lead to fractions of g smaller than 1 – they "dilute" gravity — but this system actually enhances g by a quantifiable amount.

It is tricky to design a quantifiable test of this result but the "bigger than g" part can be crudely shown. First, take a uniform rod, hold it parallel to the ground at both ends with a coin balanced on one of the ends. Let go of both ends of the rod at the same time, so it falls flat and watch the coin carefully. You will see it falls at exactly the same rate as the rod, and will even appear to stick to the rod if dropped really carefully. Now do the same but with one end of the rod resting on the edge of a table and the other held by hand, with the coin on the end away from the table. Let go now so the rod topples, rather than falls flat and watch the coin carefully. You will see it now falls behind the end of the rod (and may appear suspended in the air for a fraction of a second), thus verifying the end of the rod accelerates at a greater rate than g.

16.2.7. *Energy Approach*

We have now taken the physics of this situation quite far just using forces, but things can be taken further still using the conservation of energy. This subsection is fairly advanced but is worth a look to see just how far the analysis can be taken.

Let us consider the case for when a *uniform rod is toppled from vertical*:

If the zero of potential at the summit is defined to be zero, then the loss of height of the centre of mass on falling is $l(1 - \cos \theta)$ and so the fall in gravitational potential energy is $mgl(1 - \cos \theta)$.

The rod can be considered to undergo a combination of translation of the centre of mass at speed v_{cm} plus a rotation about the centre of mass with angular velocity ω.

The kinetic energy of the rod is then given by $\frac{1}{2}mv_{cm}^2 + \frac{1}{2}I_{cm}\omega^2$.

As the speed of the centre of mass can be written $v_{cm} = l\omega$ and moment of inertia about the centre of mass of the rod is $\frac{1}{12}m(2l)^2 = \frac{1}{3}ml^2$, the total kinetic energy can be expressed by $\frac{2}{3}ml^2\omega^2$.

Equating this with the loss in gravitational potential energy gives an expression for the angular velocity as a function of angle of $\omega^2 = \frac{3}{2}\frac{g}{l}(1 - \cos \theta)$.

This gives an expression for the tangential velocity of the centre of mass to be $\frac{v^2}{l^2} = \frac{3}{2}\frac{g}{l}(1 - \cos \theta)$, so the tangential velocity of the centre of mass is $v^2 = \frac{3}{2}gl(1 - \cos \theta)$.

16.2.8. *Variation of Forces with Angle*

There are three forces on the toppling rod which all act on the centre of mass. The velocities deduced from the energy argument allow us to find the acceleration of the centre of mass as a function of angle and hence get closer to finding the forces:

The *radial force* is given by $m\frac{v^2}{l}$ and is supplied by the components of the forces along the axis of the rod, i.e. $m\frac{v^2}{l} = mg\cos\theta - F_V\cos\theta - F_H\sin\theta$, so $\frac{3}{2}mg(1 - \cos\theta) = mg\cos\theta - F_V\cos\theta - F_H\sin\theta$, or:

$$\frac{1}{2}(5\cos\theta - 3)mg = F_V\cos\theta + F_H\sin\theta. \tag{1}$$

The *tangential force* is given by $ml\frac{d\omega}{dt} = \frac{3}{4}mg\sin\theta$ from the equation for the angular acceleration and incorporating the moment of the inertia of the rod giving $\frac{3}{4}mg\sin\theta = mg\sin\theta - F_V\sin\theta + F_H\cos\theta$, or:

$$\frac{1}{4}\sin\theta \cdot mg = F_V\sin\theta - F_H\cos\theta. \tag{2}$$

We now have two simultaneous equations, (1) and (2) that can be used to find the values of the forces for any angle of θ. After some algebra (you can try it — it requires a little care and patience, or alternatively just use a computer algebra package like Maple) the forces as a function of angle are:

$$F_H = \frac{3}{4}(3\cos\theta - 2) \cdot \sin\theta \cdot mg,$$

$$F_V = \frac{1}{4}(9\cos^2\theta - 6\cos\theta + 1) \cdot mg.$$

The plots for these forces as a function of angle are shown in Figure 16.3 for a weight of $mg = 1$N.

Analysing the forces in turn:

(1) The **weight** stays constant.
(2) The **vertical force** starts at a local maximum equal to the weight and falls steadily, reaching a minimum of zero at approximately 70°. It then rises again to a global maximum of *four times* the weight when at 180°. Then it falls down to zero at approximately 290° and rises back to unity again to complete the cycle.

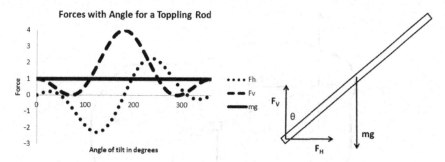

Figure 16.3: Forces as a function of angle for a uniform rod pivoted at one end when released from the vertical.

(3) The **horizontal force** starts at zero and rises, pushing the rod out to the right, reaching a peak of approximately $0.22mg$ at $30°$. It is at this point that an object toppling on the ground would be most likely to slip. The force then falls, reaching zero at around $50°$ and then becoming increasingly negative (pulling the rod's centre of mass inwards), reaching a global minimum of around $-2.3mg$ at $120°$. The force then rises, becoming zero at $180°$ and thenceforth becoming positive, pulling the rod to the right once more. It reaches a global maximum of $+2.3mg$ at $240°$ before falling to zero once more at $310°$, reaching a local minimum of $-0.22mg$ at $330°$ then rising to zero to complete the cycle.

16.2.9. *Oscillations About the Stable Equilibrium Point*

The object can be released such that it oscillates about the stable equilibrium point at an angle θ from the vertical as shown in Figure 16.4, now for a completely general shape. This type of pendulum is known as a *compound pendulum* — as distinct from the simple pendulum of Section 12.3.

In this case, using Newton's second law of motion for rotation gives $I\frac{d^2\theta}{dt^2} = -mgd\sin\theta$ where d is the distance between the pivot

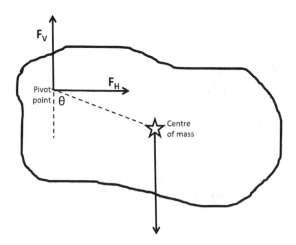

Figure 16.4:   A general compound pendulum.

point and the centre of gravity. This has no general solution but if the angle is small then $\frac{d^2\theta}{dt^2} \approx -\left(\frac{mgd}{I}\right)\theta$.

This implies the pendulum oscillates about the equilibrium point with an angular frequency of $\frac{mgd}{I}$ and period

$$T = 2\pi\sqrt{\frac{I}{mgd}}.$$

Let us test this general formula for two examples:

(1) A point mass m on the end of a light string of length l is essentially a simple pendulum. In this case, the moment of inertia is ml^2 where the distance from pivot to centre of mass is just l, so the equation simplifies to $T = 2\pi\sqrt{\frac{l}{g}}$, as one would expect. This method of finding the formula for the period of oscillation of a simple pendulum is actually more robust than the original method seen in Section 12.3 as in that case the arc that the bob moved along was approximated to a straight line.

(2) A uniform rod of length l and mass m pivoted at its end has a moment of inertia about the pivot of $\frac{1}{3}ml^2$ and a distance from

pivot to centre of mass of $\frac{1}{2}l$. The general formula for the period is now $T = 2\pi\sqrt{\frac{2}{3}\frac{l}{g}}$.

The period of oscillation of a uniform rod is thus $\sqrt{\frac{2}{3}} \approx 0.8$ times that of a simple pendulum of the same length — i.e. the rod oscillates "faster" than the string with a weight. Note that this is mass independent. This is something to be wary of when using simple pendula in experiments — if the mass is large in comparison with the length of the string, the period will be a little bit less than the ideal situation of a point mass. For better accuracy, the compound pendulum formula should be used with values of I and d included to a precision befitting the experiment.

17
Rolling and Sliding

This chapter studies objects of spherical cross section (wheel-shaped objects) that move across a surface by rolling or sliding. The condition necessary for rolling is introduced as is the idea of rolling friction. Rolling and sliding on an inclined plane is also studied.

17.1. The Condition for Rolling

Consider a solid sphere of mass m, moment of inertia about the centre of mass I and radius R that starts its motion by purely sliding (with no rotation) along a rough, hard surface. The free body diagram is shown in Figure 17.1.

Vertically, the contact force equals the weight so the sphere is in vertical equilibrium.

Horizontally, the sphere is slowed by the frictional force such that $m\frac{dv}{dt} = -F$. The velocity of the centre of mass of sphere reduces as it moves.

Rotationally, the friction puts a clockwise torque about the centre of the sphere of magnitude FR so it starts rotating in clockwise direction about the centre of mass such that $I\frac{d\omega}{dt} = FR$.

As the centre of mass *decreases* in speed and the angular velocity about the centre of mass *increases*, a time will be reached when the speed of the centre of mass relative to an observer on the floor is exactly equal to the tangential speed at the edge of the ball.

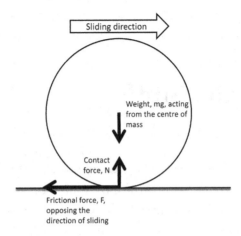

Figure 17.1:   Sphere sliding on a rough surface.

At this time, the ball has reached the *condition for rolling*. From this point onwards, the ball will no longer slip on the floor but roll without slipping such that the contact point of the rolling object has *no motion relative to the ground*.

An object that rolls without sliding at centre of mass speed v can be thought of as having a *purely translational* motion of speed v superposed on a *purely rotational* motion with angular velocity given by $\omega = \frac{v}{R}$. This is shown with the following series of kinematic illustrations in Figures 17.2a–17.2c, which show the velocity vectors at several places on the cross section of an object purely translating, purely rolling and rolling without sliding. All the figures show velocity vectors only — they are not free body diagrams as force vectors are not included.

Firstly, consider the velocity vectors at several points on an object sliding without rolling — i.e. undergoing a pure translation (Figure 17.2a).

In this case, the velocity vectors are equal in magnitude and direction at all points on the moving object.

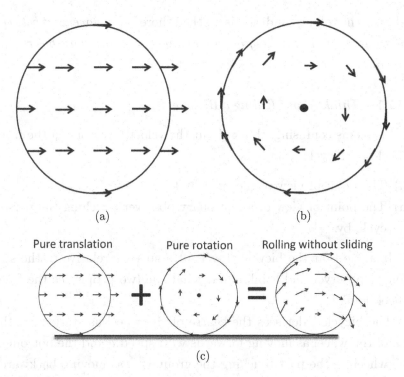

(a)   (b)

Pure translation   Pure rotation   Rolling without sliding

(c)

Figure 17.2:   (a) Velocity vectors on a sphere undergoing a pure translation — i.e. sliding without rolling.(b) Velocity vectors on a sphere undergoing a pure rotation. (c) Illustration of how rolling without slipping can be thought of as translation plus rotation.

Now consider the same object undergoing a pure rotation about its centre such that the velocity vectors at the edge of the object are of the same magnitude as the velocity of the purely translating object. The centre of the object will be stationery and the velocity vectors increase linearly with distance from the centre, as illustrated in Figure 17.2b.

When an object rolls without slipping, the velocity vectors on the rolling object are as a vector addition of the purely translating and purely rotating object, as shown in Figure 17.2c.

Rolling without slipping is a counter-intuitive concept as friction between the rolling object and the ground *must* exist for it to occur, yet once the rolling condition is reached there is *no movement between the contact point and the ground and hence the work done by friction is zero.*

17.1.1. *Think About Riding a Bicycle*

If this seems confusing, think about the velocity vectors on the front wheel of a bicycle from

(i) The point of view of the rider and,
(ii) The point of view of a stationary observer watching the person cycle by.

Imagine that the bicycle travels at a speed v relative to the stationary observer and think about what the two people see the front wheel doing.

The bicycle rider sees the centre of the wheel as stationary, the top of the wheel as moving forwards with speed v and the bottom of the wheel — the part touching the ground — as moving backwards at speed v. They also see the ground beneath the bicycle moving backwards at speed v. Kinematically, it is as if the bicycle were stationary and as they pedalled the ground moved backwards underneath them.

The stationary observer sees the rider themselves and the centre of the wheel as moving forward with speed v. They see the top of the wheel as moving forward with speed $2v$ and they see the bottom of the wheel at rest relative to the ground.

This is true regardless of the speed and acceleration of the bicycle, provided there is no slipping.

17.2. Rolling Friction — Why Rolling Objects Stop at All

If a solid object rolls without slipping on a rigid floor then there will be no resultant external force nor torque and theoretically the object will roll forever. If this idea seems silly then consider how slowly a marble loses speed when rolling across a hard table, or a bowling ball along an alley. Compared to a sliding object on most surfaces, rolling objects lose energy a very slow rate.

But rolling objects do slow down, largely due to deformations of the rolling object and the surface it rolls on. Consider two (exaggerated) cases of a wheel on a road, as illustrated in Figure 17.3.

In both cases, the leading edge of the wheel experiences a combination of contact and sliding frictional forces (that have been split into two perpendicular components on the figure) that cause the wheel to slow. The horizontal force, F_H, serves to linearly slow the wheel and the vertical force, F_V, serves to put a torque about the centre of the wheel, thus slowing the rotation. This will be bigger than the positive torque from F_H.

The greater the deformation, the greater the rolling friction. Unlike with sliding friction, rolling friction is very dependent on surface area of contact and relative velocity between ground and moving object.

Another factor that causes rolling objects to slow is adhesion between the ground and the wheel — if the wheel tries to roll over a sticky surface then it will also slow down.

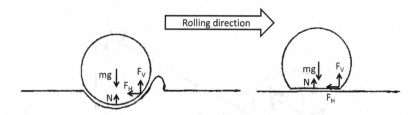

Figure 17.3: Rolling friction is caused by deformations in the wheel and the road. The left-hand image could be of a wheel rolling through sand, while the right-hand image could be of a wheel with a deflated tyre.

17.3. Rolling Down an Inclined Plane

If an object of spherical cross section rolls down an inclined plane without slipping, air resistance or rolling friction then its speed at the bottom can be determined using energy and force arguments. Before going through the analysis, try and guess the answer to the following question:

Imagine you have a sphere, a solid cylinder and a hollowed out cylinder, all of the same mass and radius. You let them roll down a hill. Which reaches the bottom of the hill first? Write down you answer, and, if you have anything suitable at your disposal, try an experiment to see which wins. Taking lumps of plasticine of the same size and making them into the shapes would be one way to do it.

17.3.1. *Analysis Using Energy*

The gravitational potential energy lost by the object in Figure 17.4 is mgH.

The linear kinetic energy gained by the object is $\frac{1}{2}mv_{CM}^2$.

The rotational kinetic energy gained by the object is $\frac{1}{2}I\omega^2 = \frac{1}{2}I\frac{v_{CM}^2}{R^2}$.

Hence, by the conservation of energy $mgH = \frac{1}{2}mv_{CM}^2 + \frac{1}{2}I\frac{v_{CM}^2}{R^2}$.

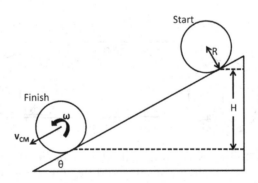

Figure 17.4: Object rolling down an inclined plane.

Rearranging gives a value for the speed of the object of:

$$v_{CM} = \left(\frac{2mgH}{m + \frac{I}{R^2}} \right)^{\frac{1}{2}} = \left(\frac{2gH}{1+k} \right)^{\frac{1}{2}} = \frac{v_P}{\sqrt{1+k}},$$

where $v_P = \sqrt{2gH}, 0 < k \leq 1$.

We are now in a position to compare the speeds of different rolling objects if we know their moment of inertia.

- For a point particle (zero moment of inertia), the speed becomes $v_P = \sqrt{2gH} \approx 1.4\sqrt{gH}$ as would be expected from the conservation of energy.
- A sphere with $I = \frac{2}{5}mR^2$ (i.e. $k = \frac{2}{5}$) reaches $\sqrt{\frac{10}{7}gH} \approx 1.2\sqrt{gH} = 0.85 v_P$.
- A solid cylinder with $I = \frac{1}{2}mR^2$ (i.e. $k = \frac{1}{2}$) reaches $\sqrt{\frac{4}{3}gH} \approx 1.1\sqrt{gH} = 0.82 v_P$.
- A thin-walled hollow cylinder with $I = mR^2$ (i.e. $k = 1$) reaches $\sqrt{gH} \approx 0.71 v_P$.

Therefore, if an object rolls down a hill, it reaches a smaller speed (and thus takes longer to reach the bottom) than an object that slides down the hill in the absence of friction. This is because in the case of sliding, all of the potential energy lost by the object goes into its translational kinetic energy. If the object rolls, the potential energy lost is shared between translational and rotational kinetic energy so it goes slower.

The more energy needs to be put into the rotation, the slower the rolling process will be. Therefore, the higher the moment of inertia, the slower the object will roll, as has been quantified above.

How well did your reasoning (and experiment if you tried it) tally with the theoretical results?

17.3.2. *Analysis Using Dynamics*

The object in Figure 17.5 accelerates along x so $ma = mg\sin\theta - F$.

The torque on the object gives an angular acceleration given by $I\frac{d\omega}{dt} = FR$ clockwise.

For rolling without slipping, $v_{CM} = R\omega$ and $a_{CM} = R\frac{d\omega}{dt}$.

This gives a value for the friction of $F = \frac{Ia}{R^2}$ and therefore $ma = mg\sin\theta - \frac{Ia}{R^2}$.

This rearranges to give a centre of mass acceleration of:

$$a = \left(\frac{m}{m + \frac{I}{R^2}}\right)g\sin\theta = \left(\frac{1}{1+k}\right)g\sin\theta = \frac{a_P}{1+k},$$

where $a_P = g\sin\theta, 0 < k \leq 1$.

This is a constant acceleration and if we wished we could use it to verify the velocity results derived by energy conservation. Instead, note that:

- For a point mass the acceleration is $a_P = g\sin\theta$, as expected;
- For a sphere it is $\frac{5}{7}g\sin\theta \approx 0.71a_P$;
- For a solid cylinder it is $\frac{2}{3}g\sin\theta \approx 0.67a_P$;
- For a hollow cylinder it is $\frac{1}{2}g\sin\theta = 0.5a_P$,

which all give qualitative agreement with the previous conclusion.

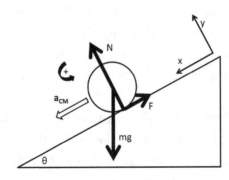

Figure 17.5: Free body diagram for an object rolling down an inclined plane.

17.3.3. *The Condition for No Slipping*

If there is no rolling then an object slides down a slope at an angle given by $\tan^{-1}\mu$. But if an object rolls, at what angle will slipping rather than rolling occur? It seems intuitive that it will probably be greater, and indeed it is, as can thus be proved:

Considering a sphere only, the value of the friction that supplies the torque to roll the ball is given by $F = \frac{2}{7}mg\sin\theta$.

If this is greater than the maximum amount of friction the contact between the surfaces can supply (μN) then the sphere will *slip* rather than roll.

As $N = mg\cos\theta$ for equilibrium in the y direction, this gives a limiting value for the angle of $\mu = \frac{F}{N} = \frac{\frac{2}{7}mg\sin\theta}{mg\cos\theta} = \frac{2}{7}\tan\theta$.

This means that if $\theta > \tan^{-1}(\frac{7}{2}\mu)$, the sphere will slip rather than roll. Notice that this is therefore always greater than the critical slipping angle for a point mass.

For example, the coefficient of friction between concrete and rubber is about 1, so if a rubber block is placed on an concrete surface at an angle then it will remain in place unless the angle exceeds $\tan^{-1}1 = 45°$. But if a rubber sphere is placed on a concrete plane it will roll without slipping unless the plane is at an angle of $\tan^{-1}\left(\frac{7}{2}\times 1\right) \approx 74°$.

17.4. An External Force Causing Rolling on a Flat Surface

This example shows how the frictional force created when pushing a rolling object forward can act in either direction to either assist *or* resist both rotational and linear acceleration. This is done by analysing the dynamics of an object being rolled forwards by a single driving force.

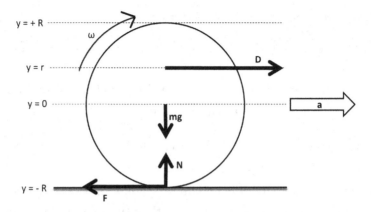

Figure 17.6:   Free body diagram for a constant force pushing a sphere.

Consider driving an object of spherical cross section, radius R, moment of inertia about the centre of mass I, mass m with a force D parallel to the ground, as illustrated in Figure 17.6.

Let the coefficient of friction between the sphere and the ground be arbitrarily large so slipping cannot occur. Of course, this is an idealised situation. In the analysis that follows, if the frictional force ever reached the limit given by $F_{max} = \mu N = \mu mg$ then the object would slip in the direction of the frictional force.

The linear acceleration of the object is given by Newton's second law, so:

$$ma = D - F. \tag{17.1}$$

The angular acceleration of the object is given by the rotational form of Newton's second law, so $I\frac{d\omega}{dt} = Dr + FR$ where r is the y-coordinate of the driving force as defined on Figure 17.6.

The no-slip condition for rolling is that $a = R\frac{d\omega}{dt}$, giving:

$$\frac{Ia}{R} = Dr + FR. \tag{17.2}$$

Equations 17.1 and 17.2 are two simultaneous equations with two unknowns:

- Eliminating the acceleration, a, from the equations gives a value for the frictional force of $F = \frac{I - mRr}{I + mR^2} D$.
- Eliminating the friction, F, from the equations gives a value for the linear acceleration of $a = \left(\frac{R + r}{I + mR^2} \right) RD$.
- The total torque is therefore given by $\Gamma = \frac{Ia}{R} = \left(\frac{R + r}{I + mR^2} \right) ID$.

It is instructive to see how these equations behave for a specific shape, for example a uniform sphere with $I = \frac{2}{5} mR^2$, which gives:

$$F_{sphere} = \frac{1}{7} \left(2 - 5\frac{r}{R} \right) D,$$

$$a_{sphere} = \frac{5}{7} \left(1 + \frac{r}{R} \right) \frac{D}{m},$$

$$\Gamma_{sphere} = \frac{2}{7} \left(R + r \right) D.$$

Plotting any of these quantities vs. r gives a straight line graph, as demonstrated in Figure 17.7.

The equations and resulting plots have several interesting features:

- For $r = R$, corresponding to driving the ball from the top of the sphere, the linear acceleration (and the torque) is a maximum.

  If there were no friction present the ball would have topspin. The presence of friction means the friction acts to *oppose* the torque and therefore *assist* the driving force.

  This is manifested by a *negative* sign in the equation (i.e. opposite to that shown in the free body diagram in Figure 17.6).
- For $r = -R$, corresponding to driving the ball from the base of the sphere, the linear acceleration and torques are both *zero*.

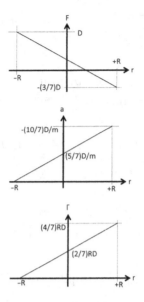

Figure 17.7:   Frictional force, acceleration, and torque with position for pushing a sphere with constant force.

The friction is a maximum in this and exactly equals the driving force.

If no friction were present, the ball would backspin.

- For $r = 0$, corresponding to striking the ball in the centre, the acceleration is at a mid level and the friction *assists* the torque and *opposes* the driving force.

If no friction were present, the ball would slide without rotating.

- For $r = \frac{2}{5}R$, corresponding to driving the ball from just above the centre, the frictional force will *always be zero*.

The driving force causes the ball to accelerate and rotate such that the linear acceleration exactly equals the transverse acceleration of the surface of the ball.

This is independent of the magnitude of the driving force; it is something that is intuitively sensed by snooker players who know that striking a ball at a certain point above the centre will minimise slipping on the table.

Remember that the specifics hold just for a uniform sphere, but the same general principles hold for any object with a circular cross section.

18

Angular Momentum

This chapter introduces angular momentum properly. It provides (or rather, repeats) the definition of angular momentum, derives two important equations relating it to torque and angular velocity, and gives a proof of an important law: the conservation of angular momentum. Several important examples of the law then follow.

18.1. Definition

(This is identical to Section 14.5.)

Consider a particle of mass m moving at velocity v at displacement r from a point O, as illustrated in Figure 18.1.

The *linear* momentum of the particle relative to O is $p = mv$ and the *angular* momentum, L, of the particle about the point is *defined* by:

$$L = r \times p = mr \times v \qquad (18.1)$$

and has magnitude $L = mrv \sin \theta$ and direction given by the right-hand rule.

Angular momentum is a vector with SI units of N·m·s or kg·m^2·s^{-1}.

18.2. Torque and Angular Momentum

If a resultant external force, F, acts on the particle in Figure 18.1 then the linear momentum of the particle must change, and therefore

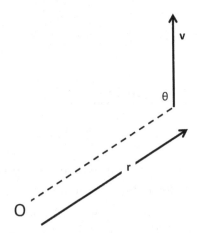

Figure 18.1: A particle moving relative to a fixed point.

its angular momentum may change. It will now be shown that the rate at which the angular momentum changes equals the torque due to the force.

From the definition of angular momentum, its rate of change is given by $\frac{dL}{dt} = \frac{d(mr \times v)}{dt}$.

Using the derivative rule for a product, this gives $\frac{dL}{dt} = m\left(\left(\frac{dr}{dt} \times v\right) + \left(r \times \frac{dv}{dt}\right)\right) = m\left((v \times v) + r \times a\right)$.

The cross product of the velocity with itself must be zero and as $F = ma$ and $\Gamma = r \times F$, it can be written:

$$\frac{dL}{dt} = r \times F = \Gamma. \tag{18.2}$$

Defined verbally:

> **The rate of change of angular momentum of a particle about a point equals the torque due to the resultant external force acting on it.**

This is the direct rotational analogy with our original form of Newton's second law (see Section 4.3), which stated that rate of change of linear momentum of a particle is directly proportional to the resulting external force acting on it.

18.3. Moment of Inertia and Angular Momentum

If a rigid (i.e. fixed shape) object is rotating about an *axis of symmetry* then the angular momentum of the object is related to the momentum of inertia about the axis and the angular velocity by a formula directly analogous to $p = mv$:

$$L = I\omega. \tag{18.3}$$

In this special case, the angular momentum and the angular velocity are aligned along the same axis. Note, however, that in general an object does *not* rotate about its axis of symmetry and Equation 18.3 is not valid. In this case, the angular momentum and angular velocity are *not* aligned.

Despite this being a specialist case of the equation, rotation about an axis of symmetry is common, and it can be used to illustrate some important angular momentum phenomena.

18.4. The Conservation of Angular Momentum

If no external torque acts on a particle, or system of particles, then Equation 18.2 states that the rate of change of angular momentum will be zero. This therefore implies that if no torque acts on a system then the angular momentum is constant. The following is therefore true:

> **If no external torques act on a system then the total angular momentum of the system about that remains constant.**

This is the *conservation of angular momentum* and (as with the conservation of linear momentum and energy) is always found to be true; there has never been an experiment performed that has refuted it. It is true at all scales from the spin of subatomic particles (where a quantum mechanical form of the law must be used) to the swirling motion of systems of galaxies. Nature provides many clear examples of which a few are presented in the next section.

18.5. Examples of the Conservation of Angular Momentum

18.5.1. *The Ice Skater (Or Less Agile Person Sat on a Rotating Platform)*

Consider a non-rigid object rotating about its centre of mass with angular velocity ω_i and moment of inertia I_i such that Equation 18.3 holds and the angular momentum of the system is given by $L = I_i\omega_i$.

If the object is able to change its shape *using internal forces alone* and maintain the axis of symmetry then the moment of inertia will, in general, be changed to a new value: I_f.

Because no external force, nor external torque, has acted on the object, its angular momentum must remain constant and therefore the angular velocity must change according to:

$$\omega_f = \frac{I_i}{I_f}\omega_i.$$

Hence, if an object's moment of inertia about the axis of rotation is reduced, then its angular velocity increases — it "spins faster". There are several examples of this in nature. In astrophysics, a rotating system of stars and planets spins faster the closer the objects get and slower the more they move apart. High divers and gymnasts are able to utilise the effect when performing aerial somersaults where a tucked figure rotates faster than an outstretched one. Ice skaters and ballerinas performing pirouettes draw their arms into their body to reduce the moment of inertia and increase the speed of rotation.

Much less skilled performers can demonstrate the effect with a rotating platform and some extra masses. If your school or university has a rotating platform that can be used for such demonstrations

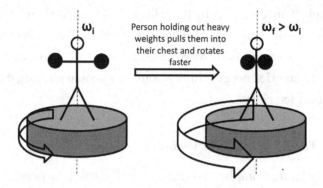

Figure 18.2: Reducing moment of inertia causes angular velocity to increase.

then you are strongly recommended to try standing on the platform with arms outstretched and spinning it slowly. Simply bringing your arms to your side causes the rotation rate to noticeably increase; holding a heavy object in each hand accentuates the effect (see Figure 18.2).

18.5.1.1. *An explanation of the kinetic energy increase*

It is interesting to consider the kinetic energy (which is entirely rotational) of the system in both states:

The initial kinetic energy is given by $\frac{1}{2}I_i\omega_i^2 = \frac{1}{2}L\omega_i = \frac{1}{2}\frac{L^2}{I_i}$ (with vector notation dropped as the angular momentum and angular velocity are along the same axis so their dot product is a maximum).

The final kinetic energy is $\frac{1}{2}I_f\omega_f^2 = \frac{1}{2}L\omega_f = \frac{1}{2}\frac{L^2}{I_f}$.

It can be seen that decreasing the moment of inertia *increases* the rotational kinetic energy and vice versa. This increase in energy is the work done *by* the person *on* the masses in pulling them in. In the reverse process, the masses would do work on the person. If the person's arms were elastic springs, the outstretched point would correspond to elastic stored energy in the spring (whereas humans burn the energy as heat).

Though it might seem to have little to do with them, this is the rotational equivalent of a *superelastic collision* (see Section 10.4) where a non-conservative force does internal work on a system to increase its kinetic energy (think about two people pushing each other apart on an ice rink).

18.5.2. *The Bicycle Wheel Variant*

A slightly more complicated variant of this is if a person of mass M on the platform is handed a rotating bicycle wheel of mass m with the wheel held parallel to the floor so its axis is aligned with the gravitational field. If the person then rotates the revolving bike wheel by 180° so it is still parallel to the floor but rotating in the opposite direction, then the person and platform start rotating in the opposite direction, thus conserving angular momentum.

A further question relates to what happens if a person stands on the Earth and tilts the wheel. What happens to the angular momentum in this case?

18.5.3. *Turning Yourself Around Without Translational Motion on An Ice Rink*

If you stand on a completely frictionless surface is it possible to turn yourself around 180° so you face the other way without grabbing on to any nearby objects to give yourself a little push?

I have tried asking this question to many students over the years, and the initial reaction from nearly everyone is "No". When asked why, the usual response is something that invokes the conservation of angular momentum.

This is, however, incorrect. It is perfectly possible, both in theory and in practice to make the turn without breaking the laws of physics. Oddly enough, when placed on a frictionless surface everyone is able to make the turn without too much thought (one could try this at an ice rink — in the lab I have a rotating platform to stand on which achieves the same result). What people tend to do is an odd looking dance of sorts where they raise their arms, swing them round and drop them down again, turning all the time.

At its simplest consider it like this: you raise one arm and hold it out in front of you. There should be no movement. You then swing it round parallel to the ground so it points out to the side. During the swing process the arm rotated in one direction, and the rest of your body rotated a little bit the other way. When the arm stopped, you stopped and angular momentum was conserved at all times. You then drop the arm back to your side and move it back to the original position. Overall you experience a net change in direction. At all times during the process your total angular momentum was zero but your angle of orientation varied.

18.5.4. *The Physics of the Falling Cat*

This method of turning around on a frictionless surface is a far cry from feline agility and balance but an appreciation of the physics of one aspect of a cat's motion follows on from this, namely that a cat always lands on its feet even when dropped feet uppermost. This appears to violate the conservation of angular momentum as a cat turns through $180°$ without a resultant external torque being applied.

The accepted theory as to how this occurs can be explained by considering the cat as being made of two halves that rotate together at angle to each other:

The cat is initially upside down and stationary so has zero linear or angular momentum. Once released, the cat arches its back so it can be thought of as being in two sections at an angle to each other. The cat rotates these sections with angular velocities ω_1 and ω_2 of the same magnitude so the total angular velocity is $\omega_{total} = \omega_1 + \omega_2$

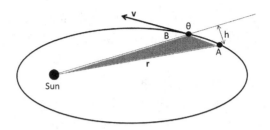

Figure 18.3: A planet moving in an elliptical path.

parallel to the ground. If the two halves of the cat have approximately the same moment of inertia then the angular momentum due to this rotation alone will also be parallel to the ground.

Since there is no external torque, the total angular momentum must always be zero. As the twisting halves of the cat create an internal angular momentum in one direction, the body as a *whole* must rotate in the opposite direction to create a net angular momentum of zero.

18.5.5. *Kepler's Second Law*

In 1609 Johannes Kepler put forward his second law of planetary motion which stated that the line joining the Sun to a planet sweeps out equal areas in equal time intervals. Later that century Newton was able to show that this was a consequence of his laws of motion plus the law of universal gravitation.

The gravitational force by the Sun on a planet is a central force — i.e. it acts along the line joining the two bodies. Consider a planet orbiting the Sun moving in an ellipse from point A to point B in a time interval δt sweeping out an approximately triangular area, as shown in Figure 18.3.

The area swept out is approximately triangular, where the triangle has a base length of r and height of h and thus an area of $\frac{1}{2}rh$.

As the torque on an orbiting planet is given by $\mathbf{\Gamma} = \mathbf{r} \times \mathbf{F}$ and the force and radius vector are aligned, the torque on the planet is zero.

As shown in Section 17.2, this implies that the angular momentum of the planet about the Sun cannot change in either direction or magnitude, hence the motion stays in the same plane with the same magnitude of $L = mrv\sin\theta$ at all positions.

In a time interval δt, the planet moves a distance $v \cdot \delta t$ from A to B. The area swept out by the planet is therefore:

$$\delta(area) = \frac{1}{2}rh = \frac{1}{2}rv \cdot \delta t \cdot \sin(180 - \theta) = \frac{1}{2}rv \cdot \delta t \cdot \sin\theta.$$

This gives a rate of area swept of $\frac{\delta(area)}{\delta t} = \frac{1}{2}rv\sin\theta$.

As the angular momentum of the planet is $mrv\sin\theta$, this implies:

$$\frac{\delta(area)}{\delta t} = \frac{1}{2}\frac{L}{m} = constant.$$

Kepler's second law is thus mathematically verified.

An interesting feature of this proof is that it has at no point invoked the inverse square law of gravity — it works equally well for any *central force* that is a function of distance from the origin only.

19

Angular Momentum, Gyroscopes, and Precession

This chapter introduces gyroscopic motion and the phenomenon of torque-induced precession. While the mathematics used is not particularly sophisticated, the central phenomenon is rather counterintuitive and conceptually tricky. The most basic case of gyroscopic motion, with precession in the plane of the ground is looked at in detail; more subtle aspects of gyroscopic motion are mentioned with the necessary physics briefly outlined.

To develop a good understanding of the phenomena in this section, it is strongly recommended that you obtain a toy gyroscope and experiment with it to see how the phenomena work in real life as well as following the theoretical analysis.

19.1. The Gyroscope

A simple gyroscope is a flywheel (a freely rotating wheel that does not roll but simply stores rotational kinetic energy) that is allowed to spin freely about a horizontal axis with one end placed on a pivot, as illustrated in Figure 19.1.

If the flywheel is spun fast enough and one end of the gyro is placed on the pivot then instead of toppling off (as it would do if it were not spinning) the other end moves in a circular motion about the pivot. As the angular velocity of the gyro's flywheel slows down (by friction) the circular motion about the pivot speeds up. Even if

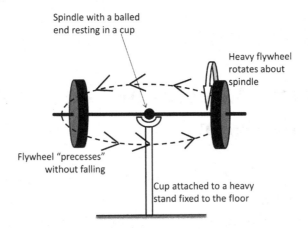

Spindle with a balled
end resting in a cup

Heavy flywheel
rotates about
spindle

Flywheel "precesses"
without falling

Cup attached to a heavy
stand fixed to the floor

Figure 19.1:   Sketch of a gyroscope precessing in a plane about a pivot.

you do not have a gyroscope to try this with, a similar effect can be
seen with a spinning top or any object that can be kept upright by
spinning it fast on its end when it would otherwise topple (e.g. a coin
at a slight angle to the vertical).

This phenomenon is known as *precession* and will be analysed in
three ways:

(1) By consideration of the angular momentum vectors and torque
    by gravity about the pivot,
(2) By analogy with linear circular motion, and,
(3) By consideration of forces and velocities.

19.2.   Application of Torque about the Pivot to a Spinning Gyro

One almost trivial case is considered first before looking at the case
of actual precession.

(a) Downward torque with no initial angular momentum

Consider the case of placing the gyro on the pivot when it <u>is not</u>
spinning, as illustrated in Figure 19.2.

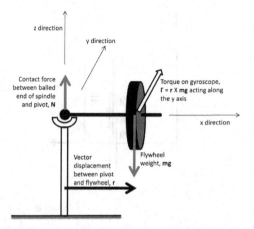

Figure 19.2: **The forces and torque on a gyroscope with a stationary flywheel.**

Assume all the weight of the gyro is in the flywheel (i.e. the rod placed along the x-axis is very light in comparison with the flywheel so that the torque on the system is in the y-direction). Torque is rate of change of angular momentum, so in time δt the *change* in angular momentum is $\delta \boldsymbol{L} = \boldsymbol{\Gamma} \cdot \delta t = \boldsymbol{r} \times m\boldsymbol{g} \cdot \delta t$.

With each additional time increment, the angular momentum increments will add in the y-direction as the gyro topples faster and faster. If viewed from above (along the z-axis as denoted in Figure 19.2) then a vector diagram for the angular momentum increase can be represented as shown in Figure 19.3.

(b) Torque perpendicular to initial angular momentum

Consider now the case when the flywheel of the gyro is already spinning so it has an initial angular momentum, and the gyro is placed on the pivot so there is a torque that acts *perpendicular* to the initial angular momentum, as shown in Figure 19.4.

The *change* in angular momentum when the gyroscope is released will be the same as in part (a) with no initial spinning — i.e. an amount $\delta \boldsymbol{L} = \boldsymbol{\Gamma} \cdot \delta t = \boldsymbol{r} \times m\boldsymbol{g} \cdot \delta t$ in the y-direction.

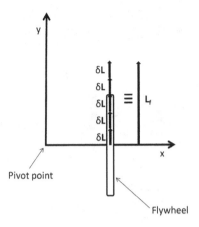

Figure 19.3: Incremental increase in angular momentum for a toppling flywheel.

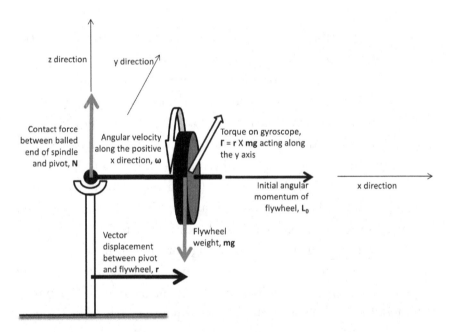

Figure 19.4: Forces and torque on a gyroscope with a rotating flywheel.

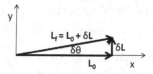

Figure 19.5: The change in angular momentum occurring solely in the *xy*-plane for an incremental time step.

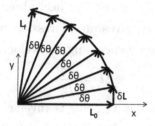

Figure 19.6: The precession of a gyroscope in the plane due to an addition of successive small changes in angular momentum in the plane.

To get the new angular momentum, this must be added vectorially to the original angular momentum — for an infinitesimally small time increment this means the new momentum will *not have changed in magnitude, but will have changed in direction.* The view from above is shown in Figure 19.5.

Thus, with each small time increment, the gyro changes its orientation but not its magnitude of angular momentum and hence moves about the pivot in a circle, as illustrated in Figure 19.6.

The flywheel thus moves in a circular motion about the pivot. The necessary force required for this circular motion comes from a combination of friction and contact forces by the pivot on the end of the gyroscope's connecting rod.

19.3. Precession Formula

For steady precession in a plane, a simple expression can be found for the magnitude of the precession angular velocity, Ω.

In time δt the gyroscope's axis turns an angle $\delta\theta$ which can be seen to be equivalent to the magnitude of ratio $\frac{\delta L}{L}$ from Figure 19.6. As $\Omega = \frac{d\theta}{dt}$ by definition, this gives $\Omega = \frac{\frac{dL}{L}}{dt} = \frac{1}{L}\frac{dL}{dt} = \frac{\Gamma}{L}$ where Γ is the magnitude of the torque about the pivot as shown in Figure 18.5 and can be written mgr.

Furthermore, as the gyro precesses about an axis of symmetry, if the moment of inertia of the flywheel about the axis of the gyro is I then $L = I\omega$, which gives an expression relating the precession angular velocity to the flywheel angular velocity thus:

$$\Omega = \frac{mgr}{I\omega}. \tag{19.1}$$

The precession angular velocity is inversely proportional to the flywheel's angular velocity. This explains why the precession angular velocity increases as frictional energy losses occur.

Although Equation 19.1 has been derived for the special case of the gyro precessing in the xy-plane, it can be proved (see, e.g. *Newtonian Mechanics* by A.P. French, 1971) that the formula holds for any gyro angle to the vertical.[1]

19.4.   Analogy with Linear Circular Motion

Precession is a rotational analogue of uniform circular motion: a particle attached to a string initially at rest and pulled with a force towards a person obviously moves towards the person. But if the particle is moving and the string is pulled perpendicular to the motion, the particle will tend to move in a circle — the *change* in momentum is still towards the person, but the *new* momentum is the same magnitude but in a different direction. With the gyro the same argument applies, but with angular momentum and torque in place of linear momentum and force.

---

[1] Actually a very mild dependence on angle is predicted — though the author has never seen this experimentally verified.

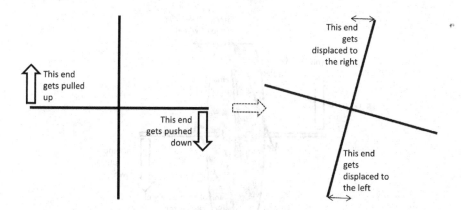

Figure 19.7: Twisting two perpendicular rods about a pivot.

19.5. Analysis of Precession in Terms of Forces and Velocities

Two perpendicular connected rods with equal and opposite forces applied to the ends of one of the rods see the endpoints of the other rods undergo displacements perpendicular to the direction of the forces, as illustrated in Figure 19.7.

Hence, consider applying a torque to a rotating bicycle flywheel and consider the *change* in velocities of the top and bottom of the wheel, as shown in Figure 19.8.

The forces at the ends of the spindle cause a change in the plane of rotation such that the wheel shifts *perpendicular* to the applied forces. This leads to the same conclusion as the angular momentum analysis.

19.6. Precession is Nothing to do with the Conservation of Angular Momentum

One final point to make on the phenomenon of precession is effectively summed up by the title of this section, but it is such a common misconception that is does merit a short discussion. Note that the

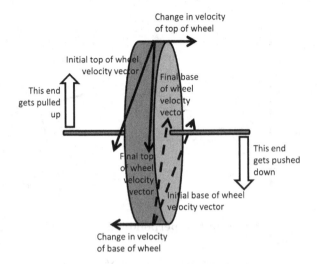

Figure 19.8: **The velocity change at the rim of a rotating wheel subject to twisting forces.**

conservation of angular momentum only applies to objects with no resultant external torque upon them. Gyroscopes do have a resultant external torque upon them — indeed this is the agent that is responsible for the precession in the first place and angular momentum is most certainly not conserved. It is unfortunately frequent for students of the subject to invoke angular momentum conservation in the description of the behaviour of gyroscopes.

That said, there is a subtlety here in that in the z-direction as noted in Figures 19.2–19.6, there is no external torque. How about angular momentum along that axis? When a gyroscope is initially placed on the pivot, it is not precessing and therefore the angular momentum along the z-axis is exactly zero. But once the gyroscope is precessing then angular momentum along z is most definitely non-zero even though there is no torque in that direction. How so? To help answer this either take a toy gyroscope or if you don't have one watch a video and make a careful observation of the motion as the flywheel slows down.

19.7. More Subtle Features of Gyroscopic Motion

When the gyroscope is freed it actually falls a little bit at first, then oscillates about two points, often steadying to a flat precession in a plane. This oscillation is known as **nutation**.

In fact, the final precession level is below the original position from which the gyroscope is freed. This loss of gravitational potential energy supplies the necessary energy for the precessional rotation. The downward pointing of the gyroscope also ensures angular momentum is conserved in the z-direction which answers the questions at the end of the previous subsection.

For an expanded and more formal analysis see:

- *The Feynman Lectures on Physics, Volume 1*, Chapter 20, for a good descriptive version;
- French, *Newtonian Mechanics*, Chapter 14, for a clear and detailed mathematical treatment;
- McCall, *Classical Mechanics*, Chapter 8
- Fowles and Cassiday, *Analytical Mechanics*, Chapter 9.

19.8. The Earth's Precession

The Earth is not a uniform sphere, and the axis of rotation is inclined slightly to the poles' axis of symmetry. As a result, the Earth undergoes two types of natural precession:

(1) "Free" precession (which has not been discussed here) as a natural wobble (as when any spinning object wobbles). The period is about 305 days.
(2) Precession caused by torques applied by the Sun and the Moon. The period of this is about 26,000 years.

19.9.   Examples and Uses of Gyroscopic Motion

As gyroscopes tend to resist changes in orientation, they are valuable and widely used devices in navigation and engineering. This is probably the most frequent practical use of gyroscopes though they are often not physically visible and hidden within sealed containers.

With an understanding of the basics of precessions there are various other phenomena that can be explained using the same physics. The closing challenge for the reader is to consider the following phenomena and how the physics of precession applies:

- A coin is difficult to balance on its edge when stationary but easy to balance when rolling.
- Similarly, a bicycle is easy to keep upright when moving but difficult when stationary.
- The barrel of some firearms are grooved to cause the projectile exiting the barrel to exit spinning about its long axis (known as "rifling" and hence the name "rifle").
- Similarly, when throwing an American football in the style of a quarterback, it is usually necessary to spin the ball.
- High-speed motorcycle cornering cannot be achieved by large turns on the handlebars and most of the turning effects are achieved by leaning and weight shifting by the rider.
- In turning left or right in a helicopter, it is tricky for beginner pilots to maintain a constant altitude.
- In tenpin bowling, expert players are able to achieve more "hook" with a throw by using a bowling ball with a non-symmetrical internal density distribution.

Bibliography

Einstein, A. (1905). Zur elektrodynamik bewegter Körper, *Annalen der physic* 17 (10) pp. 891–921. English translation available from http://www.fourmilab. ch/etexts/einstein/specrel/www/ [Accessed 16 June 2015].

Feynman, R.P., Leighton, R.B. and Sands, M. (1963). *The Feynman Lectures on Physics, Volume 1*, Boston: Addison-Wesley Publishing Co.

French, A.P. (1971). *Newtonian Mechanics (The M.I.T. Introductory Physics Series)*, New York: W.W. Norton & Company.

McCall, M.W. (2011). *Classical Mechanics: From Newton to Einstein: A Modern Introduction*, (2nd edition), Chichester: John Wiley & Sons.

Newton, I., Bouguer, M.P. and Cavendish, H. (1900). *The Laws of Gravitation: Memoirs of Newton, Bouguer and Cavendish, Together with Abstracts of Other Important Memoirs*, New York: American Book Company.

Ogborn, J. and Taylor, E.F. (2005). Quantum physics explains Newton's laws of motion, *Physics Education* **40** (1), 26–34.

Young, H.D. and Freedman, R.A. (2006). *University Physics with Modern Physics*, (12th edition), Boston: Addison-Wesley Publishing Co.

Index

Printed in the United States
By Bookmasters